T0139753

Advanced Structured Materials

Volume 119

Series Editors

Andreas Öchsner, Faculty of Mechanical Engineering, Esslingen University of
Applied Sciences, Esslingen, Germany
Lucas F. M. da Silva, Department of Mechanical Engineering, Faculty of
Engineering, University of Porto, Porto, Portugal
Holm Altenbach, Faculty of Mechanical Engineering,
Otto-von-Guericke-Universität Magdeburg, Magdeburg, Sachsen-Anhalt, Germany

Common engineering materials reach in many applications their limits and new developments are required to fulfil increasing demands on engineering materials. The performance of materials can be increased by combining different materials to achieve better properties than a single constituent or by shaping the material or constituents in a specific structure. The interaction between material and structure may arise on different length scales, such as micro-, meso- or macroscale, and offers possible applications in quite diverse fields.

This book series addresses the fundamental relationship between materials and their structure on the overall properties (e.g. mechanical, thermal, chemical or magnetic etc) and applications.

The topics of *Advanced Structured Materials* include but are not limited to

- classical fibre-reinforced composites (e.g. glass, carbon or Aramid reinforced plastics)
- metal matrix composites (MMCs)
- micro porous composites
- micro channel materials
- multilayered materials
- cellular materials (e.g., metallic or polymer foams, sponges, hollow sphere structures)
- porous materials
- truss structures
- nanocomposite materials
- biomaterials
- nanoporous metals
- concrete
- coated materials
- smart materials

Advanced Structured Materials is indexed in Google Scholar and Scopus.

More information about this series at http://www.springer.com/series/8611

Muhamad Husaini Abu Bakar ·
Mohamad Sabri Mohamad Sidik ·
Andreas Öchsner

Editors

Progress in Engineering Technology

Automotive, Energy Generation, Quality Control and Efficiency

 Springer

Editors
Muhamad Husaini Abu Bakar
Malaysian Spanish Institute
Universiti Kuala Lumpur
Kulim, Kedah, Malaysia

Mohamad Sabri Mohamad Sidik
Malaysian Spanish Institute
Universiti Kuala Lumpur
Kulim, Kedah, Malaysia

Andreas Öchsner
Faculty of Mechanical Engineering
Esslingen University of Applied Sciences
Esslingen, Baden-Württemberg, Germany

ISSN 1869-8433   ISSN 1869-8441   (electronic)
Advanced Structured Materials
ISBN 978-3-030-28507-4   ISBN 978-3-030-28505-0   (eBook)
https://doi.org/10.1007/978-3-030-28505-0

This Springer imprint is published by the registered company Springer Nature Switzerland AG
The registered company address is: Gewerbestrasse 11, 6330 Cham, Switzerland

Preface

This book contains the selected and peer-reviewed manuscripts that were presented during the Conference on Language, Education, Engineering and Technology (COLEET 2018), held at the University Kuala Lumpur Malaysian Spanish Institute (UniKL MSI) from November 13 to 14, 2018. COLEET 2018 is an annual international conference aimed at presenting current and ongoing research being carried out in the fields of mechanical, manufacturing, electrical, and electronics engineering technology. This volume provides in-depth ongoing research activities among academia of UniKL MSI, and it is hoped to foster cooperation among organizations and researchers involved in the covered fields.

Kulim, Malaysia   Muhamad Husaini Abu Bakar
Kulim, Malaysia   Mohamad Sabri Mohamad Sidik
Esslingen, Germany   Andreas Öchsner

Contents

Study the Effect of Acetone as an Inhibitor for the Performance of Aluminium-Air Batteries

Mohamad-Syafiq Mohd-Kamal, Muhamad Husaini Abu Bakar and Sazali Yaacob

Abstract Aluminium-air battery have high energy density, for example 8100 Wh kg^{-1} capable of replacing classical lithium based batteries. However, the presence of parasitic reactions during the discharge process causes reducing the lifetime of the aluminium-air battery. Organic inhibitors are able to prevent the parasitic reaction, but it is likely to effect the battery performance. The aim of this research is to study the effect of acetone as an inhibitor at aluminium-air battery. Density functional theory (DFT) with B3LYP functional and 6-311G(d,p) basis set was conducted to determine the inhibitor efficiency of acetone. Besides, the aluminium-air battery was developed and tested to identify battery performances by applying acetone with different concentrations (0, 2, 4, 6, and 8 mM). Results show that increasing the acetone concentration will improve the inhibitor's efficiency from 12.5 to 50.0%. Further, the capacity of the battery can be increased with the inhibitor concentration. It is observed that the battery capacity using acetone (8 mM) is 0.028 Ah better than for a battery without acetone, 0.023 Ah. Therefore, acetone can be considered as an inhibitor capable of preventing severe corrosion against aluminium alloys and produces a good performance of aluminium-air batteries.

Keywords Aluminium-air battery · Acetone derivatives · Inhibition efficiency · DFT · Battery performance

M.-S. Mohd-Kamal · M. H. Abu Bakar (✉) · S. Yaacob
System Engineering and Energy Laboratory, Universiti Kuala Lumpur
Malaysian Spanish Institute, Kulim Hi-Tech Park, 09000 Kulim, Kedah, Malaysia
e-mail: muhamadhusaini@unikl.edu.my

M.-S. Mohd-Kamal
e-mail: msyafiq.kamal@s.unikl.edu.my

S. Yaacob
e-mail: sazali.yaacob@unikl.edu.my

1 Introduction

In recent years, metal-air batteries have become an attraction for battery replacement technology, as it offers many advantages [1–3]. The metal-air battery acts by producing an electrochemical energy conversion that allows the chemical energy of the metal to be converted into electrical energy [4, 5]. Moreover, the metal-air battery has many types of material anodes, and the most attractive candidate is aluminium [6–8]. The aluminium-air battery has in a theory of high energy density of 8100 Wh kg^{-1} [3, 9].

The aluminium used as an anode electrode for aluminium-air batteries proved to be effective with low atomic weight, low toxicity, low cost and high power (2980 Ah kg^{-1}). Aluminium can be extracted from abundant sources and is accessible to discover. However, for this aluminium-air battery, self-corrosion will occur on the surface of the aluminium electrode [7, 10]. Corrosion caused by parasitic reactions leads to a reduction in the lifetime of this aluminium-air battery. The reduced efficiency of the energy performance from this parasitic reaction makes the commercialization of aluminium-air batteries difficult [11].

Several investigations have been proposed to solve the problem of the parasitic reaction that occurs in this aluminium electrode, and the best method is to use an inhibitor in the battery [12–14]. In previous studies conducted by Nie et al. [10], the addition of organic compounds as inhibitors of an electrolyte solution can help to reduce the corrosion of the parasitic reactions of the aluminium electrodes. The organic inhibitors can act as activators of the dissolution of the aluminium electrodes and do not stop the activity of the aluminium electrodes [10, 15]. It has been shown that corrosion inhibitors consisting of acetone are effective in reducing the corrosion of aluminium by forming a stable barrier layer [16, 17].

Generally, inhibitors with O or N atoms can produce a good barrier, but if the inhibitor comprises both atoms is better [18]. The molecules of the inhibitors will interact with the corrosion reactions of aluminium, and these molecules can block the surface of the corrosive agent [19, 20]. The performance of these organic inhibitors will depend on the electronic structure, mechanical properties, donor density, molecular area, molecular weight of the inhibitor and the chemical properties of the adsorption coating formed on the metal surfaces [21, 22].

In this study, aluminium-air batteries are fabricating and tested to analyze the difference in battery life without the inhibitor and the dissolved battery inhibitor. Acetone with molecular properties capable of preventing corrosion on the metal surface is used as an inhibitor in this study. The acetone will dissolve in the battery electrolyte to see the ability of this inhibitor to inhibit parasitic reactions on the surface of the battery's aluminium electrode.

2 Experimental

2.1 Computational Study

Density functional theory (DFT) was used to obtain the molecules of acetone to predict the energy molecular orbital [21, 23]. Combination of Becke three-parameter hybrid (B3) exchange functional with the Lee-Yang-Parr (LYP) (B3LYP) as correction functional and 6-311G(d,p) basis-set was used in DFT to determine the HOMO-LUMO energy for acetone [24, 25]. Figure 1 below shows the acetone structure that was used as inhibitor in aluminium-air battery.

   Furthermore, HOMO-LUMO orbital was used to calculated the energy gap, electron affinity (EA) and ionization potential (IP) [26]. The value of the gap energy is calculated using, E gap = ELUMO-EHOMO, the energy difference implies low reactivity of the chemical species when the value is higher [27, 28]. The IP and EA was calculated from the energy of HOMO and LUMO, respectively, within the framework of Koopmans' theorem [29]:

$$IP = -E_{HOMO} \tag{1}$$

$$EA = -E_{LUMO} \tag{2}$$

where E_{HOMO} is the energy of HOMO and E_{LUMO} is the energy of the LUMO, respectively.

2.2 Materials

In this experiment, there are three important components to build aluminium-air batteries which is an anode, cathode, and electrolyte [11, 30]. Figure 2a shows the 8 mm × 6.5 mm × 1.5 mm of aluminium alloy A1100 were used as an anodes [5]. The aluminium alloy has the following chemical compounds (% by weight) at 99.5%, Cu 0.2%, Fe 0.95%, Mn 0.05%, Si 0.95% and Zn 0.1% [31].

   Figure 2b shows the 8 mm × 6.5 mm × 1 mm of the air cathode which was done by binding the iron mesh and the activated carbon. The activated carbon was produced by the pyrolysis process, i.e. by was immersing in 2 M potassium hydroxide for 24 h [32]. The air cathode used in the aluminium-air battery act as catalyst energy to battery [33–35].

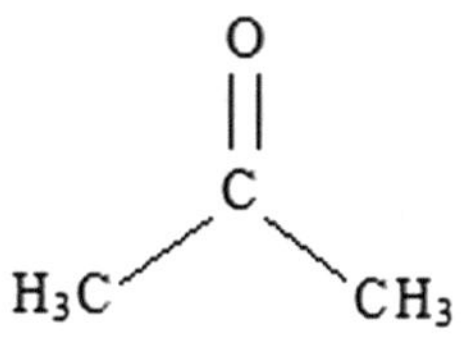

Fig. 1 The structure of acetone

Fig. 2 **a** Aluminium alloy used as an electrode. **b** Air cathode used in the aluminium-air battery. **c** The parasitic reaction during battery discharge

The electrolytes used in this experiment were 1 M sodium hydroxide [11]. Several concentration of acetone (2, 4, 6, and 8 mM) were used to as inhibitor [16, 26]. The different concentration of acetone dissolved in NaOH were used to show the effect to the anode.

2.3   Weight Loss

Inhibition efficiency measurement was performed using the method of weight loss of aluminium during battery test [19, 36, 37]. The measurement test focused on the specimen (anode) by determining the weight (g). The initial weight of the specimen was taken before inserting it into the battery. The battery was tested for one hour. After one hour, the specimen was removed from the battery, washed with water, dried and weighed.

The weight loss of the specimens was calculated from the initial and final weight. The experiment was repeated to see the concentration of inhibitors in the 1 M NaOH solution and acetone (2, 4, 6, 8 mM). This procedure shows the difference in weight loss of the specimen for an electrolyte that has different concentration of the inhibitor.

The efficiency of the inhibitor is determined using the following relationship [38]:

$$IE(\%) = \frac{W_o - W_i}{W_i} \times 100 \tag{3}$$

where W_o is the weight loss without inhibitor and W_i is the weight loss with inhibitor.

2.4 Battery Test

The aluminium-air batteries were developed with aluminium anodes, air cathodes, and electrolyte NaOH. An aluminium-air battery test was performed using the Arduino battery capacity tester at a constant resistance load of 10 Ω as shown in Fig. 3. The aluminium-air batteries was being tested to see the lifetime capable of the battery capability for one cell. This test distinguishes between batteries with different concentration inhibitors.

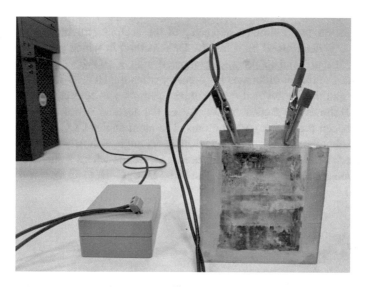

Fig. 3 Experiment setup for Aluminium-air battery

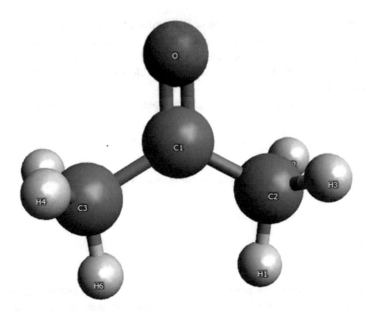

Fig. 4 Optimized geometry of acetone structure

3 Results and Discussion

3.1 Optimised Geometry

Figure 4 shows the optimized geometry of the acetone structure. The optimized geometry was determined by using the DFT method in which parameters such as bond length and bond angle were compared after a HOMO-LUMO calculation. There is no significant difference in geometry when the HOMO-LUMO calculation was done and this shows that the optimized geometry is correct.

Table 1 shows the angles and bonds for the acetone structure which have 15 angles between the vertex carbon atoms. The atom angles CCC, CCO, HCH, and CCH represent the optimized angles where every angle is mostly correct. Moreover, there were 9 bonds optimized the for acetone structure where 8 single bonds and a double bond are shown in Table 1. The Bond H–C, C–H, C–C represent the single bonds and C=O is the double bond.

3.2 HOMO-LUMO Energy

Molecular orbital is describing the space where the probability of found electron is high which can use to determine the inhibitor efficiency [39]. HOMO tends to donate electrons to the other molecule that has less electron. LUMO is likely to be

Table 1 Bonds and angles for the acetone molecule after optimization

	Type	Start atom	Vertex	End atom	Angle (°)	Length (Å)
Angle 1	CCC	C2	C1	C3	117.0408	–
Angle 2	CCO	C2	C1	O	122.0749	–
Angle 3	CCO	C3	C1	O	120.8843	–
Angle 4	HCH	H1	C2	H3	109.8777	–
Angle 5	CCH	C1	C2	H1	110.1693	–
Angle 6	HCH	H1	C2	H2	106.7050	–
Angle 7	CCH	C1	C2	H3	109.9702	–
Angle 8	HCH	H2	C2	H3	109.8736	–
Angle 9	CCH	C1	C2	H2	110.1930	–
Angle 10	HCH	H4	C3	H6	109.3171	–
Angle 11	CCH	C1	C3	H4	108.8197	–
Angle 12	HCH	H4	C3	H5	107.0489	–
Angle 13	CCH	C1	C3	H6	113.3525	–
Angle 14	HCH	H5	C3	H6	109.2665	–
Angle 15	CCH	C1	C3	H5	108.8425	–
Bond 1	H–C	H1	–	C2	–	1.09536
Bond 2	H–C	H4	–	C3	–	1.09329
Bond 3	H–C	H6	–	C3	–	1.09121
Bond 4	C–H	C2	–	H3	–	0.08915
Bond 5	C–C	C2	–	C1	–	1.51602
Bond 6	C–H	C2	–	H2	–	1.09557
Bond 7	C–C	C3	–	C1	–	1.52419
Bond 8	C–H	C3	–	H5	–	1.09361
Bond 9	C=O	C1	–	O	–	1.20882

the molecule whose orbital has fewer electrons and indicates the potential to receive the electron from HOMO.

Figure 5 shows the acetone HOMO-LUMO orbital which were determined by using DFT. The blue orbital shown in Fig. 2 is the positive base for the electron movement, while red is for negative base. The HOMO orbital is the molecule that easily donates the electrons to the LUMO orbital. The Orbital in HOMO illustrates the movement of electrons that are donated and the LUMO orbital shows the best area to accept electrons. The electron donor HOMO and the electron acceptor LUMO produce the energy gap. The energy gap determines the electron conductivity in acetone and characterizes the acetone molecular chemical reactivity.

As shown in Fig. 2, the HOMO and LUMO energies determined are −6.861 and −0.535 eV, respectively. Therefore, the energy gap is 6.326 eV. The reactivity of acetone depends on the energy gap, where the energy gap is high, then the reactivity of acetone is low. In addition, HOMO-LUMO orbital were used to estimate the

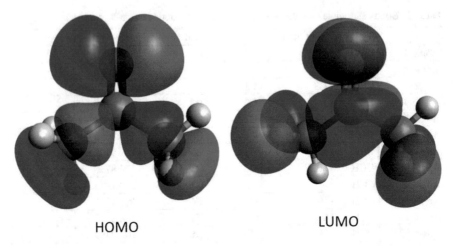

HOMO   LUMO

Fig. 5 The HOMO-LUMO orbital of the molecule acetone

value of IP and EA. The HOMO energy to be the negative of the IP of the acetone and the EA is known as accepting electrons. The value of IP and EA are 6.861 and 0.535 eV, respectively.

3.3 Inhibition Efficiency

Aluminium hydroxide caused by the corrosion of aluminium and the reactions of electrolytes in the battery effects the performance of the cell [40]. The formation of an uncontrolled oxide film that occurs in the battery contributes to decreasing the lifetime of the battery. Reduction of the corrosion of metal in an electrolyte solution is difficult due to the specific addition of OH to the metal surface [41, 42]. Mainar et al. states that the coagulation process occurs very challenging in the dissolution of metal in an alkaline electrolyte that caused all the organic matter to be rejected by coating the layer. Therefore, the acetone was synthesized by determining the inhibitor efficiency of acetone.

Table 2 shows the results of weight loss of aluminium alloy (anode) by several concentrations during battery testing. Anode without acetone shows the highest

Table 2 The values of inhibitor efficiency, weight loss, and inhibitor efficiency

Inhibitor	Concentration (mM)	Weight loss (g)	Inhibitor efficiency (%)
Blank	–	0.18	–
Acetone	2	0.16	12.5
	4	0.14	28.57
	6	0.13	38.46
	8	0.11	50.0

weight loss which is 0.18 g. The corrosion happening in aluminium is higher due to OH which Cn easily reach the aluminium surface. The weight loss of aluminium was decreased when 2 mM of acetone was added in the electrolyte battery which is 0.16 g. Even though the value of weight loss decreased is small, but it can have improved the inhibitor efficiency of 12.5% and control corrosion.

A 4 mM of acetone increased the inhibitor efficiency by a value of 28.57% and reduced the weight loss of the aluminium anode. Acetone produced atomic O to block the OH from contact with the aluminium anode. This is shown in Fig. 6 where O was absorbed into the aluminium alloy. The value of weight loss for an acetone concentration of 6 mM was decreased to 0.13 g. The inhibitor efficiency was depending on the reduction of weight loss, where the inhibitor efficiency for 6 mM was increased to 38.46%. The highest inhibitor efficiency was 50.0% from acetone with 8 mM. Acetone worked very well in reducing corrosion with weight loss for aluminium reduced to 0.11 g.

Figure 7 shows the value of the inhibitor efficiency resulting from the concentration of acetone in the battery's electrolyte solution. This efficiency value indicates the agility of the acetone molecule capable of the controlling the erosion that occurs in aluminium alloys. The efficiency value of inhibitor was shown with different concentration capable of reducing the corrosion of acetone in aluminium. As a result of this study, it can be observed that the concentration of acetone plays an important role in the control of corrosion. The increase in the efficiency of the inhibitor from 12.5 to 50.0% showed the ability of the acetone which acted on the parasitic reactions that occur in the anode.

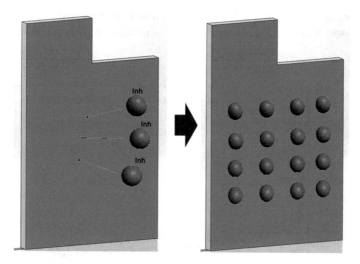

Fig. 6 Atomic O acting as an inhibitor to aluminium alloy

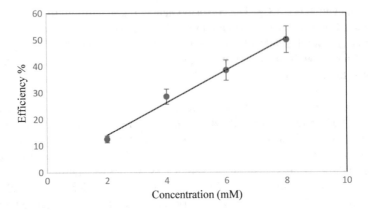

Fig. 7 Inhibitor efficiency against concentration

3.4 Voltage Discharge Rate

The performance voltage of the aluminium-air battery can change over time. The initial voltage of the aluminium-air battery can reach more than 1.3 V at room temperature. This experiment demonstrated the ability of acetone as inhibitors to improve the performance of the battery through the voltage discharge rate. Figure 8 shows the voltage performance at each concentration value that dissolves in the electrolyte.

Experiments that take one hour for each different concentration represent slightly similar voltage performance. It is considered that acetone can control parasitic reactions, and is able to maintain the battery performance. This is shown in Fig. 8,

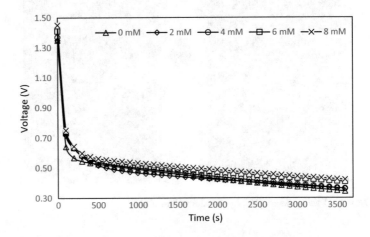

Fig. 8 Performance of the aluminium-air battery according to the concentration inhibitor value

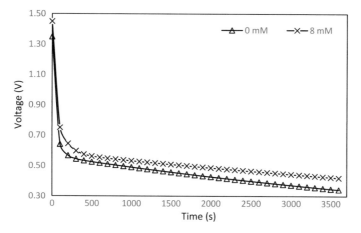

Fig. 9 Discharge voltage rate performance for a battery with an inhibitor concentration of 8 mM and without inhibitor

the voltage at the beginning of each concentration reaches more than 1.3 V. The voltage in the batteries without acetone as an inhibitor has the same value as the inhibitor batteries, although the corrosion in the aluminium alloys is higher than that of the aluminium alloys that are supplied. Moreover, the battery with an 8 mM concentration of acetone shows the highest trend of voltage discharge rate. Therefore, the higher concentration of acetone can increase the value of the inhibitor efficiency while maintaining the battery performance. However, it is also possible that when the concentration value is too high, the battery performance will decrease because the reaction between electrolyte and aluminium is not strong. Therefore, the optimal concentration for the inhibitor should be emphasized.

An 8 mM concentration of acetone was selected to test the lifetime of the aluminium-air battery. Figure 9 shows the difference voltage discharge rate between a battery with acetone (8 mM) and without acetone. Both cells were tested for 1 h, and it turns out that both values show the same trend, which is the characteristic of aluminium-air batteries.

The graph indicates that the initial voltage is 1.35 V for the battery without inhibitor. The graph shows a drastic declined in early stage of discharge until reached 0.5 V. After that, the voltage continuous discharge slowly until reached a voltage at 0.3 V as depicted in Fig. 9 within 1 h. The aluminium-air battery with acetone (8 mM) shows the graph of voltage discharge almost the same as the battery without the acetone. The maximum voltage of this battery is 1.45 V. A drastically discharge voltage is shown in the graph above, whereat the voltage is decreased from 1.45 to 0.6 V. Then, the discharge voltage slowly drops to 0.4 V within 1 h. The performance of the battery voltage with acetone is better than for the battery without acetone.

3.5 Capacity Discharge Rate

The battery capacity is a significant role in the production of battery discharge performance. The effective capacity of the battery needed performance is twice the value of the capacity within a few hours for highly rated applications such as electric vehicles. The performance of this capacity battery must be achieved for the use at high power. However, the low power capability is also able to offer good launch performance to smaller electronic applications. Therefore, this experiment examined the performance of the aluminium-air battery capacity by looking at the effectiveness of the acetone as an inhibitor on battery capacity.

Figure 10 presents the capacity of the aluminium-air battery with several concentrations of acetone during 1 h testing. The capacity of the battery was reaching more than 0.02 Ah. The capacity trend of the battery shows increasing.

Table 3 shows the specific values of capacity of the battery within 1 h. Battery without acetone represent the value capacity 0.023 Ah. The capacity battery with 2 mM of acetone demonstrate some improved at which the maximum capacity is 0.024 Ah. As shown, there are differenced capacities between the battery without acetone and the battery with acetone. The capacity was increased with increasing the concentration of acetone. An 8 mM acetone battery shows the highest capacity

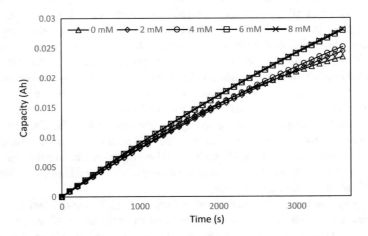

Fig. 10 The performance of capacity battery with several concentrations of acetone for 1 h

Table 3 Values of acetone concentration and capacity of the battery	Inhibitor	Concentration (mM)	Capacity (Ah) after 1 h
	Blank	–	0.023
	Acetone	2	0.024
		4	0.025
		6	0.027
		8	0.028

performance at which the value capacity is 0.028 Ah. Therefore, the acetone is capable of maintaining and improving the capacity of the aluminium-air battery.

4 Conclusions

Several concentrations of acetone (2, 4, 6, and 8 mM) were used as a corrosion inhibitor to solve the problem of the parasitic reaction that occurs in the aluminium electrode. The acetone can act as activators of the dissolution of the aluminium electrodes and control the reaction of the anode and electrolyte. In this study, aluminium-air batteries were developed and tested to analyze the difference in battery performance with acetone and without acetone. The increase in the efficiency of the inhibitor from 12.5 to 50.0% showed the ability of the acetone molecule acting on the parasitic reactions. Furthermore, the capacity performance of the battery without inhibitor shows the value at 0.023 Ah, but a battery with the inhibitor (8 mM) has 0.028 Ah. The capacity value of the battery that has an inhibitor is better than the battery without inhibitor. Thus, can conclude that acetone can be used as an inhibitor that is capable of preventing severe corrosion against aluminium alloys. Moreover, the selected acetone concentration is essential for the electrolyte battery solution to produce a good chemical reaction.

Acknowledgements The authors would like to thank the STRG program [grant number str17065] and are grateful for the support given from the Yayasan Tengku Abdullah Scholarship (YTAS) under Universiti Kuala Lumpur and System Engineering and Energy Laboratory (SEELab).

References

1. Gelman, D., et al.: An aluminum—ionic liquid interface sustaining a durable Al-air battery. J. Power Sources **364**, 110–120 (2017)
2. Pino, M., et al.: Performance of commercial aluminium alloys as anodes in gelled electrolyte aluminium-air batteries. J. Power Sources **299**, 195–201 (2015)
3. Liu, Y., et al.: A comprehensive review on recent progress in aluminum-air batteries. Green Energy Environ. **2**, 246–277 (2017)
4. Moghadam, Z., et al.: Electrochemical performance of aluminium alloy in strong alkaline media by urea and thiourea as inhibitor for aluminium-air batteries. J. Mol. Liq. **242**, 971–978 (2017)
5. Park, I.J., et al.: Aluminum anode for aluminum-air battery—Part II: influence of In addition on the electrochemical characteristics of Al–Zn alloy in alkaline solution. J. Power Sources **357**, 47–55 (2017)
6. Fouda, A.S., Mohamed, N.H.: Corrosion inhibition of aluminum in hydrochloric acid solutions using some. Int. J. Electrochem. Sci. **3**, 9861–9875 (2014)
7. Cho, Y.J., et al.: Aluminum anode for aluminum-air battery—Part I: influence of aluminum purity. J. Power Sources **277**, 370–378 (2015)

8. Yasakau, K.A., Zheludkevich, M.L., Ferreira, M.G.S,: Corrosion and Corrosion Protection of Aluminum Alloys. Elsevier (2017)
9. Han, B., Liang, G.: Neutral electrolyte aluminum air battery with open configuration. Rare Met. **25**, 360–363 (2006)
10. Nie, Y., et al.: An effective hybrid organic/inorganic inhibitor for alkaline aluminum-air fuel cells. Electrochim. Acta **248**, 478–485 (2017)
11. Avoundjian, A., Galvan, V., Gomez, F.A.: An inexpensive paper-based aluminum-air battery. Micromachines **8**, 222 (2017)
12. Kharitonov, D.S., Kurilo, I.I., Zharskii, I.M.: Effect of sodium vanadate on corrosion of AD31 aluminum alloy in acid media. Russ. J. Appl. Chem. **90**, 1089–1097 (2017)
13. Camila, G., Alexandre, F.: Corrosion Inhibitors—Principles, Mechanisms and Applications. Developments in Corrosion Protection (2014)
14. Al-Sodani, K.A.A., et al.: Efficiency of corrosion inhibitors in mitigating corrosion of steel under elevated temperature and chloride concentration. Constr. Build. Mater. **163**, 97–112 (2018)
15. Goyal, M., et al.: Organic corrosion inhibitors for industrial cleaning of ferrous and non-ferrous metals in acidic solutions: a review. J. Mol. Liq. **256**, 565–573 (2018)
16. Saratha, R., Meenakshi, R.: Dimethylaminobenzylidene acetone as corrosion inhibitor for mild steel in acid medium. Rasayan J. Chem. **4**, 251–263 (2011)
17. Yadav, M., et al.: Substituted imidazoles as corrosion inhibitors for N80 steel in hydrochloric acid. Indian J. Chem. Technol. **20**, 363–370 (2013)
18. Wang, H., et al.: DFT study of new bipyrazole derivatives and their potential activity as corrosion inhibitors. J. Mol. Model. **13**, 147–153 (2007)
19. Raghavendra, N., Ishwara, B.J.: Inhibition of Al corrosion in 0.5 M HCl solution by Areca flower extract. J. King Saud Univ. Eng. Sci. (2018)
20. Fateh, A., et al.: Review of corrosive environments for copper and its corrosion inhibitors. Arab. J. Chem. (2017)
21. Obot, I.B., et al.: Density functional theory (DFT) as a powerful tool for designing new organic corrosion inhibitors: Part 1: an overview. Corros. Sci. **99**, 1–30 (2015)
22. Obayes, H.R., et al.: Sulphonamides as corrosion inhibitor: experimental and DFT studies. J. Mol. Struct. **1138**, 27–34 (2017)
23. Wang, Y., et al.: Density functional theory analysis of structural and electronic properties of orthorhombic perovskite $CH_3 NH_3 PbI_3$. Phys. Chem. Chem. Phys. **16**, 1424–1429 (2014)
24. Toy, M., Tanak, H.: Molecular structure and vibrational and chemical shift assignments of 3'-chloro-4-dimethylamino azobenzene by DFT calculations. Spectrochim. Acta Part A Mol. Biomol. Spectrosc. **152**, 530–536 (2016)
25. Efil, K., Obot, I.B.: Quantum chemical investigation of the relationship between molecular structure and corrosion inhibition efficiency of benzotriazole and its alkyl-derivatives on iron. Prot. Met. Phys. Chem. Surf. **53**, 1139–1140 (2017)
26. Zarrouk, A., et al.: A theoretical study on the inhibition efficiencies of some quinoxalines as corrosion inhibitors of copper in nitric acid. J. Saudi Chem. Soc. **18**, 450–455 (2014)
27. Rostami, Z., et al.: DFT results against experimental data for electronic properties of C60 and C70 fullerene derivatives. J. Mol. Graph. Model. **81**, 60–67 (2018)
28. Kim, K.H., et al.: Basis set effects on relative energies and HOMO-LUMO energy gaps of fullerene C36. Theor. Chem. Acc. **113**, 233–237 (2005)
29. El Mahdy, A.M., et al.: DFT and TD-DFT calculations of metallotetraphenylporphyrin and metallotetraphenylporphyrin fullerene complexes as potential dye sensitizers for solar cells. J. Mol. Struct. **1160**, 415–427 (2018)
30. Mokhtar, M., et al.: Recent developments in materials for aluminum-air batteries: a review. J. Ind. Eng. Chem. **32**, 1–20 (2015)
31. Davis, J.R.: Aluminum and aluminum alloys. Light Met. Alloy. **66** (2018)
32. Montes, V., Hill, J.M.: Activated carbon production: recycling KOH to minimize waste. Mater. Lett. **220**, 238–240 (2018)

33. Zhang, E., et al.: Durability and regeneration of activated carbon air-cathodes in long-term operated microbial fuel cells. J. Power Sources **360**, 21–27 (2017)
34. Cheng, S., Wu, J.: Air-cathode preparation with activated carbon as catalyst, PTFE as binder and nickel foam as current collector for microbial fuel cells. Bioelectrochemistry **92**, 22–26 (2013)
35. Yang, W., et al.: A simple method for preparing a binder-free paper-based air cathode for microbial fuel cells. Bioresour. Technol. **241** (2017)
36. Hu, K., et al.: Influence of biomacromolecule DNA corrosion inhibitor on carbon steel. Corros. Sci. **125**, 68–76 (2017)
37. Sardar, N., Ali, H.: A study of some new acidizing inhibitors on corrosion of N-80 alloy in 15% boiling hydrochloric acid. Corrosion **58**, 317–321 (2002)
38. Ayati, N.S., et al.: Inhibitive effect of synthesized 2-(3-pyridyl)-3,4-dihydro-4-quinazolinone as a corrosion inhibitor for mild steel in hydrochloric acid. Mater. Chem. Phys. **126**, 873–879 (2011)
39. Balachandran, V., et al.: Conformational stability, spectroscopic (FT-IR & FT-Raman), HOMO-LUMO, NBO and thermodynamic function of 4-(trifloromethoxy) phenol. Spectrochim. Acta Part A Mol. Biomol. Spectrosc. **130**, 367–375 (2014)
40. Egan, D.R., et al.: Developments in electrode materials and electrolytes for aluminiumeair batteries. J. Power Sources **236**, 293–310 (2013)
41. Bösing, I., et al.: Electrolyte Composition for distinguishing corrosion mechanisms in steel alloy screening. Int. J. Corros. (2017)
42. Mainar, A.R., et al.: An overview of progress in electrolytes for secondary zinc-air batteries and other storage systems based on zinc. J. Energy Storage **15**, 304–328 (2018)

Performance Characteristics of Palm Oil Diesel Blends in a Diesel Engine

Shahril Nizam Mohamed Soid, Mohamad Ariff Subri, Mohammad Izzuddin Ariffen and Intan Shafinaz Abd. Razak

Abstract Continuous usage and excavation of crude petroleum create a global alarm that in several years, the petroleum resources would be depleted. Researches on biomass resources such as palm oil, to replace diesel fuel have been conducted comprehensively to find the most suitable alternative fuel to diesel fuel. Blends of 20, 40, 60, 80 and 100% of palm oil (PO) with diesel were investigated in a single cylinder diesel engine. Each blend was tested on a standard Yanmar 178F diesel engine at various engine speeds (1000–1800 rpm) and various loads (200–1000 kW). The performance and fuel consumption of the engine were analysed and compared. Experimental results show that performance and fuel consumptions of the engine may produce better, equal or deteriorate when running on palm oil diesel blends.

Keywords Biomass · Palm oil · Diesel fuel · Diesel engine

1 Introduction

Biomass is an energy resource derived from organic materials such as plant and animal that can generate power and electricity by using various methods. The renewable and sustainable characteristic of biomass fuel is the most desired char-

S. N. Mohamed Soid (✉) · M. A. Subri · M. I. Ariffen
Mechanical Section, Universiti Kuala Lumpur Malaysian Spanish Institute, Kulim Hi-Tech Park, 09000 Kulim, Kedah, Malaysia
e-mail: shahrilnizam@unikl.edu.my

M. A. Subri
e-mail: ariffsubri@gmail.com

M. I. Ariffen
e-mail: izzukun@gmail.com

I. S. Abd. Razak
Electrical Engineering Department, Tuanku Sultanah Bahiyah Polytechnic, Kulim Hi-Tech Park, 09000 Kulim, Kedah, Malaysia
e-mail: shafinas@ptsb.edu.my

© Springer Nature Switzerland AG 2019
M. H. Abu Bakar et al. (eds.), *Progress in Engineering Technology*,
Advanced Structured Materials 119, https://doi.org/10.1007/978-3-030-28505-0_2

acteristic that researchers consider to replace the usage of hydrocarbon or fossil based fuel. There are many types of methods in preparing biomass fuel such as fermentation conversion, compose, thermal conversion and chemical conversion. Other than that, biomass fuel derived from plant such vegetable oil also can be used in internal combustion engines.

An experiment on diesel engine using vegetable oil was conducted a decade ago. Rudolph Diesel tested his original diesel engine with peanut oil in the year of 1900. The usage of vegetable oil in diesel engine was reduced a few years later, due to the abundance and cheap supply of petroleum product [1]. Usage of peanut oil is one of the earliest proof that a diesel engine is able to run with vegetable oil, and this leads to long research for the alternative fuel.

Nowadays, many researchers do extend research to alternative fuels, mainly using vegetable oil either pure or blends of vegetable oil with diesel. This research was conducted mainly due to the growing concern towards environmental and greenhouse effects that are getting worst day by day [2]. Furthermore, vegetable oil resources also abundance in many types such as peanut oil, olive oil, sesame oil, palm oil and many more. Naturally abundance makes the study of the alternative fuel using vegetable oil more reasonable and giving a high impact for the humankind in the future.

There are several advantages of vegetable oil compared to conventional diesel fuel. The advantages of vegetable oil are easily acquired, renewable and environmentally friendly. Besides, there are still many types of vegetable oil that have a huge potential to be discovered. Vegetable oil can be renewed by replantation of the plant, and vegetable oil seldom contains sulphur, aromatic hydrocarbon, metals or crude oil residues that can lead to environmental pollution [2, 3]. One of the promising and high potential vegetable oil is palm oil.

Palm oil or their scientific name *Cocos Nucifera* is one of the most promising vegetable oil that can be acquired easily in a tropical country such as Malaysia. In Malaysia, palm oil is one of the most important export commodity for the Malaysian economy and is mainly used for cooking and food manufacturing. Other than its main role in food industries, palm oil is also able to replace diesel fuel as an alternative in diesel engines [3].

Pure palm oil is able to run on a diesel engine, but faces several degradations on performance due to several problems. Problems such as the properties of palm oil that are different from the diesel that may affect the components of the engine that relate to the performance of the engine. This study was conducted to investigate the performance of the diesel engine fuelled by palm oil and its blends with diesel with specific ratios of palm oil to diesel in percentages.

2 Literature Review

Vegetable oil can be divided into two categories that are edible and non-edible. Previous research focused more on the vegetable oil that is mainly and easily acquired from the respected country. Examples of successful experiments on non-edible vegetable oils are using putranjiva oil, jatropha oil and karanja oil, while soybean oil and palm oil were examples of successful research that run using edible vegetable oils. A typical vegetable oil structure can be seen as in Fig. 1. R^1, R^2 and R^3 are the alkyl group representation [4].

Palm oil was completely miscible to the diesel fuel, thus making it suitable for blending in any proportion of palm oil or diesel fuel [5]. Palm oil blends were created by mixing the palm oil with diesel fuel with a certain percentage of palm oil and diesel fuel. Both mixtures have about the same microstructure and were expected not to react to each other. Biodiesel fuel blends, vegetable oil blends and palm oil blends can be assumed to behave like that because both of the biodiesel and diesel fuel were non-polar, miscible to each other and add up to volume when blended to each other [5].

The difference between palm oil and diesel fuel properties can be observed from the higher density, viscosity and lower heating value of palm oil compared to diesel fuel. Palm oil possesses higher viscosity compared to diesel fuel. The high viscosity of palm oil might create a poor atomization process of the fuel that leads to an incomplete combustion, choking to the fuel injector, carbonization of piston ring and accumulation of combustion fuel into the lubrication oil [2]. Moreover, it also makes the diesel engine difficult to start at a low temperature and increases the penetration of the fuel spray that can lead to the thickening of the lubrication oil [3].

Blending of diesel fuel with various quantities of vegetable oil can reduce the viscosity of the vegetable oil. This technique can minimize the cost of fuel processing and engine modification [2]. Preheating the vegetable oil before running on diesel engine also can reduce the viscosity of the vegetable oil and avoid the fuel

Fig. 1 Vegetable oil structure

$$
\begin{array}{c}
\quad\quad\quad O \\
\quad\quad\quad \| \\
CH_2\!-\!O\!-\!C\!-\!R^1 \\
\quad\quad\quad O \\
\quad\quad\quad \| \\
CH\!-\!O\!-\!C\!-\!R^2 \\
\quad\quad\quad O \\
\quad\quad\quad \| \\
CH_2\!-\!O\!-\!C\!-\!R^3
\end{array}
$$

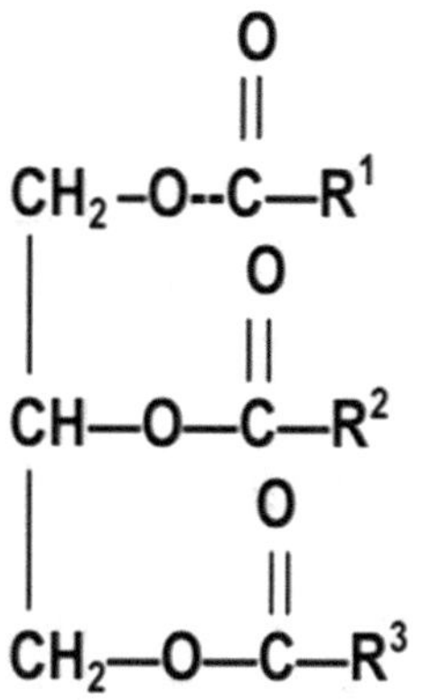

filter from clogging [6, 7]. The degumming process also helps on reducing the viscosity of the straight vegetable oil [6].

Gross heating value or gross calorific value is the measure of the total energy released when a substance is completely burned or combusted under the presence of oxygen and standard condition. Vegetable oil possesses almost the same gross heating value or gross calorific value compared to the diesel fuel. Diesel fuel has the highest gross calorific value with the lowest viscosity. Due to almost the same gross heating values, that makes vegetable oil the most suitable alternative substitute to the diesel fuel [2]. The lower the percentage of vegetable oil on vegetable oil blends, the lower the viscosity and the higher gross calorific value of the vegetable oil blends [2]. Both diesel fuel and vegetable oil can be mixed together to obtain the satisfied viscosity and the gross calorific value [3].

A diesel engine that runs with vegetable oil might be a promising future as an alternative substitution even though the performance of the engine might be lower when compared running with diesel fuel. Therefore, the usage of palm oil blends in a diesel engine should be investigated to understand its characteristics and provide a better way in improving the engine performance.

3   Experimental Set-up and Procedures

The experiments were conducted by testing a diesel engine fuelled by palm oil and its blends to gather the engine performance data such as engine torque, power, brake specific fuel consumption (BSFC), brake mean effective pressure (BMEP) and thermal efficiency. The data were recorded to obtain the performance curve for palm oil blends and compared to engine running with diesel alone.

3.1   The Engine

A Yanmar 178F compression ignition diesel engine was used in the experiment and was mounted on the test rig. The engine was coupled with a dynamometer by using the belting system. The engine power was rated at 4 kW at the maximum speed of 3600 rpm. The details about the engine specification are shown in Table 1. The experimental setup used in this study is shown in Fig. 2.

3.2   Palm Oil Diesel Blends

Palm oil-diesel blends were produced by mixing palm oil with diesel with specific quantities. Five different blends of palm oil were used in these experiments. The details of the blends are shown in Table 2.

Table 1 Specification of the compression ignition (CI) engine

Model name	Yanmar 178F
Type	Single cylinder, vertical, 4 stroke, air-cooled, direct injection
Bore × stroke	7.8 cm × 6.2 cm
Displacement	296 cubic centimeter
Compression ratio	20:1
Rated power/speed	4 kW at 3600 rpm
Rotation direction (view from wheel)	Clockwise
Nozzle pressure	200 kgf/cm^2
Fuel tank capacity	3.5 L

Fig. 2 Yanmar 178F diesel engine coupled with engine dynamometer

Table 2 Five difference mixture of palm oil-diesel blends and its density

Fuel blends	Composition	Density(g/ml)
Diesel	100% Diesel	0.828
20%PO	20% Palm oil + 80% diesel	0.8386
40%PO	40% Palm oil + 60% diesel	0.8632
60%PO	60% Palm oil + 40% diesel	0.8784
80%PO	80% Palm oil + 20% diesel	0.8896
100%PO	100% Palm oil	0.9046

3.3 Experimental Flow

The engine was run on diesel and five palm-oil blends to gather the performance data for comparison purpose. After that, the data were analyzed and compared to observe the performance of engine based on quantities such as torque, brake power, BSFC, BMEP and η_{bt}. Performance of the engine was measured by a dynamometer connected to the engine shaft. The speed of the engine was measured using a tachometer. The torque of the engine was measured using a load cell attached to the dynamometer while the fuel consumption of the engine was measured using a burette. The formulae used to analyze the performance of the engine are as follows.

$$BP = 2\pi NT \tag{1}$$

where N is the engine speed in rps and T is the torque produced by the engine. Brake mean effective pressure can be calculated using Eq. (2),

$$BMEP = \frac{2BP}{ALNn} \tag{2}$$

where A is the bore area, L is the stroke length, N is the engine speed in rps and n is the number of the cylinder. Brake thermal efficiency can be calculated using Eq. (3),

$$\eta_{bt} = \frac{BP}{\dot{m}_f \cdot Q_{net}} \tag{3}$$

where $\dot{m}_f$ is the mass flow rate of the fuel, while Q_{net} is the calorific value of the fuel when completely combusted. Brake specific fuel consumption can be calculated using Eq. (4),

$$BSFC = \frac{\dot{m}_f}{BP} \tag{4}$$

where $\dot{m}_f$ is the mass flow rate of the fuel, while BP is the brake power produced by the engine.

4 Results and Discussion

4.1 Performance of the Engine on Diesel Fuel and Palm Oil Blends

Figure 3 shows the engine torque for diesel and blends fuel throughout the load at 1000 rpm. From Fig. 3 it was found that all blends had shown similar

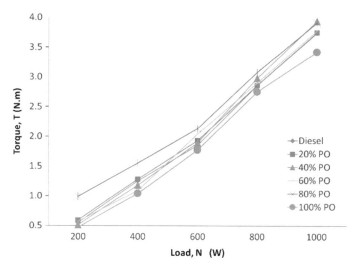

Fig. 3 Palm oil diesel blends engine torque at 1000 rpm and various engine loads

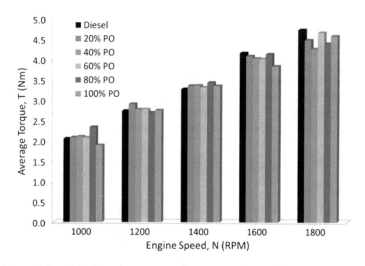

Fig. 4 Palm oil diesel blends engine average torque at a various engine speeds

characteristics as diesel. The torque of the engine was increased when the load was increased. At 1000 rpm, the lowest engine torque was produced by 100% palm oil. Surprisingly 80% of the palm oil-diesel blend had produced the highest torque among all blends. The average engine torque at various engine speeds is shown in Fig. 4. From this Fig. 4, it was found that the palm oil blends are able to produce a better engine torque compared to diesel at 1000–1400 rpm. However, beyond 1400 rpm, the palm oil blends produced lower engine torque compared to diesel alone.

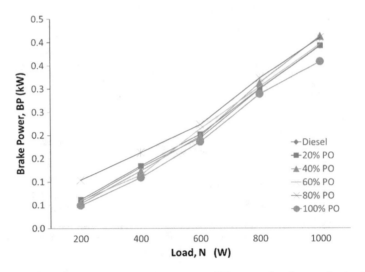

Fig. 5 Palm oil diesel blends engine brake power at 1000 rpm and various engine loads

Figure 5 shows the engine brake power comparison to diesel at 1000 rpm. Once again, all blends had shown the same characteristic as diesel and the trends show that the brake power increased when the load increased. Similar to engine torque, 80% of the palm oil-diesel blend had produced the highest brake power. The lowest brake power was produced by 100% palm oil. On average brake power analysis at various engine speeds (Fig. 6), it was found that all the palm oil-diesel blended fuels are able to produce a brake power almost equal or higher than diesel fuel at low speed (1000–1400 rpm). But it starts to decline after 1400 rpm.

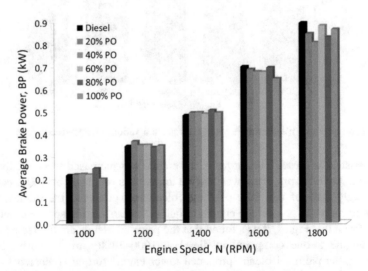

Fig. 6 Palm oil diesel blends engine average brake power at various engine speeds

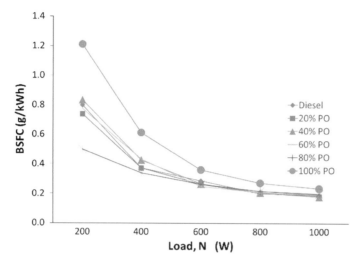

Fig. 7 Palm oil diesel blends engine BSFC at 1000 rpm and various engine loads

Figure 7 shows the brake specific fuel consumption (BSFC) comparison between palm oil blends and diesel at 1000 pm. Both diesel and palm oil blends had shown the same characteristics where the value was decreasing with increasing engine load. The engine performance is better when it has the lowest value of BSFC. From Fig. 7, it is clear that 80% of palm oil-diesel blend has the lowest value of BSFC when compared to diesel and other blends. While 100% palm oil had produced the highest BSFC. Figure 8 shows the average BSFC at various engine speeds and it was found that except for 100% palm oil, all blends had

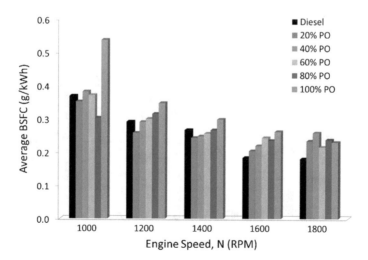

Fig. 8 Palm oil diesel blends engine average BSFC power at various engine speeds

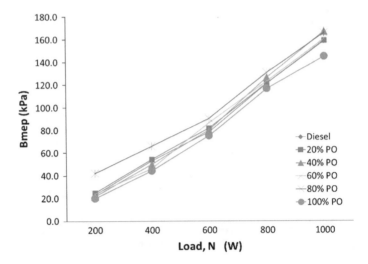

Fig. 9 Palm oil diesel blends engine bmep at 1000 rpm and various engine loads

produced almost equal or lower BSFC compared to diesel fuel at low engine speed. But after 1400 rpm, all blends had produced higher BSFC than diesel. This is due to its lower value of energy density when compared to diesel fuel.

The comparison for brake mean effective pressure between palm oil blends and diesel at 1000 rpm is shown in Fig. 9. The bmep of the engine for all fuels were increased when the load increased. The 80% palm oil diesel blend has the highest bmep compared to the other blends, while 100% palm oil has the lowest bmep. The average bmep (Fig. 10) at various engine speeds shows small differences between

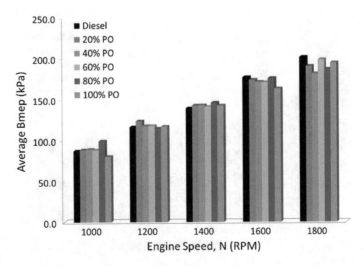

Fig. 10 Palm oil diesel blends engine average bmep power at various engine speeds

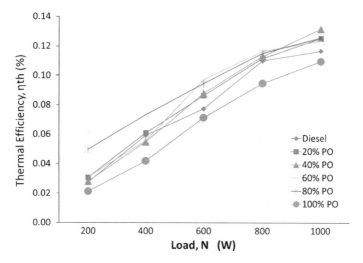

Fig. 11 Palm oil diesel blends engine brake thermal efficiency at 1000 rpm and various engine loads

all blend when compared to diesel, except for 100% palm oil. However, after 1400 rpm, the bmep from palm oil diesel blends decreased when compared to diesel alone.

The overall engine performance can be indicated by using the brake thermal efficiency, η_{bt}. Brake thermal efficiency is the ratio between the brake power and indicated power. Highest thermal efficiency is desired for engine performance. From the experiments it was found that the engine efficiency increases as the load increases at 1000 rpm as shown in Fig. 11 for diesel and all blends. Once again, the 80% palm oil diesel blend has the highest thermal efficiency when compared to others. However, at higher engine speed (above 1400 rpm), the palm oil blends thermal efficiency is lower when compared to diesel due to its low brake power and high amount of fuel consumed for the combustion process as shown in Fig. 12.

5 Conclusion

The experimental setup was configured well and the experiment was conducted successfully without any major problem during the experimental run. The diesel engine has proven to be able to run on palm oil and its blends and the performance for each blend were comparable to diesel fuel. Some fuel blends seem better or worse than diesel fuel. From the experiments, it was found that in the range of 1000–1400 rpm the performance of the palm oil blends was about the same when compared to diesel fuel. The performance curves were close to each other at all loads in term of engine torque, brake power, bsfc, bmep and thermal efficiency.

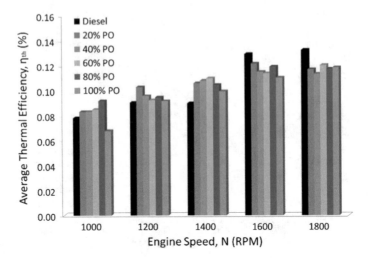

Fig. 12 Palm oil diesel blends engine average brake thermal efficiency power at various engine speeds

However at higher speed, which is more than 1400 rpm, all palm oil blends performance produce low engine performance most probably due to poor mixture quality of fuel and air before the combustion process. Therefore further investigation and improvement on the injection process is required to improve the engine performance when palm oil blends are used in diesel engine.

Acknowledgements The authors would like to thank Universiti Kuala Lumpur Malaysian Spanish Institute for the financial support for this research work via UniKL STRG grant UniKL/CoRI/strl5028.

References

1. Dermibas, A.: Biodiesel from vegetable oils via transesterification in supercritical methanol. Energy Convers. Manage. **43**, 2349–2356 (2002)
2. Wang, Y.D., Al-Shemmeri, T., Eames, P., et al.: An experimental investigation of the performance and gaseous exhaust emissions of a diesel engine using blends of vegetable oil. Appl. Therm. Eng. **26**, 1684–1691 (2006)
3. Almeida, S.C.A.D., Belchior, C.R., Nascimento, M.V.G., et al.: Performance of a diesel generator fueled with palm oil. Fuel **81**, 2097–2102 (2002)
4. Agarwal, D., Kumar, L., Agarwal, A.K.: Performance evaluation of a vegetable oil fueled compression ignition engine. Renew. Energy **33**, 1147–1156 (2008)
5. Benjumea, P., Agudelo, J., Agudelo, A.: Basic properties of palm oil biodiesel—diesel blends. Fuel **87**, 2069–2075 (2008)

6. Haldar, S.K., Ghosh, B.B., Nag, A.: Studies on the comparison of performance and emission characteristics of a diesel engine using three degummed non-edible vegetable oils. Biomass Bioenergy **33**, 1013–1018 (2009)
7. Bari, S., Lim, T.H., Yu, C.W.: Effect of preheating of crude palm oil (CPO) on injection system, performance and emission of a diesel engine. Renew. Energy **27**, 339–351 (2002)

Optimization of Palm Oil Diesel Blends Engine Performance Based on Injection Pressures and Timing

Shahril Nizam Mohamed Soid, Mohamad Ariff Subri,
Muhammad-Najib Abdul-Hamid, Mohd Riduan Ibrahim
and Muhammad Iqbal Ahmad

Abstract The usage of palm oil diesel blends as an alternative fuel in a diesel engine has been proven by many researchers. However, the high viscosity of palm oil produced heavy low-volatility compounds that are difficult to combust in the main combustion phase, and produce longer combustion period when compared to diesel. This phenomenon will contribute to poor engine performance. Therefore, this research was conducted to optimize the engine performance based on different injection pressures and timings. Blends of 20 to 100% of palm oil (PO) and diesel were investigated on a standard Yanmar 178F diesel engine at various engine speeds, loads, injection pressures and timings. The optimization was carried out by using a design of experiment software. Experimental results show that the blends can match the diesel engine performance at higher injection pressure.

Keywords Palm oil · Diesel fuel · Diesel engine · Performance optimization

S. N. Mohamed Soid (✉) · M. A. Subri · M.-N. Abdul-Hamid · M. R. Ibrahim
Mechanical Section, Universiti Kuala Lumpur Malaysian Spanish Institute, Kulim Hi-Tech
Park, 09000 Kulim, Kedah, Malaysia
e-mail: shahrilnizam@unikl.edu.my

M. A. Subri
e-mail: ariffsubri@gmail.com

M.-N. Abdul-Hamid
e-mail: mnajib@unikl.edu.my

M. R. Ibrahim
e-mail: mohdriduan@unikl.edu.my

M. I. Ahmad
Faculty of Bioengineering and Technology, Universiti Malaysia Kelantan, 17600 Jeli,
Kelantan, Malaysia
e-mail: iqbal.a@umk.edu.my

© Springer Nature Switzerland AG 2019
M. H. Abu Bakar et al. (eds.), *Progress in Engineering Technology*,
Advanced Structured Materials 119, https://doi.org/10.1007/978-3-030-28505-0_3

31

1 Introduction

Biomass is a renewable energy sources for fuel that can be developed from an organic material to create electricity or another form of energy. It is renewable and sustainable and can be obtained from various resources such as vegetable oils and animal products [1]. The usage of vegetable oil has an advantage compared to conventional diesel such as easier to produce and renewable. There are many types of vegetable oil, for examples sunflower, corn, coconut, sesame and palm oil. Palm oil is the most productive oil crops in the world, and can be the best option for biomass fuel.

Cocos nucifera or palm oil is easy to acquire in a tropical country like Malaysia. In fact, Malaysia is the second largest exporter of the palm oil behind Indonesia with the market share of 32% of the world market in 2015. With the European Union (EU) resolution and its proposal to ban palm oil into Europe, the usage of palm oil blends in Diesel engine may be on the right track in Malaysia.

The usage of palm oil diesel blends as an alternative fuel in a diesel engine has been proven by many researchers. However, its usage contributes to degradations of engine performance due to its different density and viscosity. There are a few experiments and studies that were conducted to study the effects of palm oil usage in a diesel engine.

The heavy low-volatility compounds that are produced by the high viscosity of palm oil are difficult to combust in the main combustion phase and produce a longer combustion period when compared to diesel. Injection timing and injection pressure are the parameters that can be changed to improve the combustion process. Injection timing is the duration of the injected fuel in the combustion chamber. While, injection pressure is how much the pressure of the fuel is injected into the combustion chamber that will affect the behaviour of the sprayed fuel. This study was conducted to optimize the engine performance based on different injection pressures and timings.

2 Literature Review

Pure palm oil may be employed in diesel engines as an alternative fuel. Engine performance and emissions were influenced by basic differences between diesel fuel and palm oils such as mass-based heating values, viscosity, density and molecular oxygen content. The high viscosity of palm oil resulted in poor atomisation, carbon deposits, clogging of fuel lines and starting difficulties in low temperatures [2].

Several problems can be observed when the vegetable oil is run in a diesel engine that affects the operation and durability. The example operational problems are starting the engine, ignition, combustion and performance of the diesel engine. While durability problems are related to the formation of the carbon deposits, carbonization, sticking of the oil ring and dilution of the lubrication oil [3].

Vegetable oil can make the fuel filter to choke due to long operation due to the high viscosity of the pure vegetable oil and insoluble substance that presents vegetable oil. The high viscosity of the pure vegetable oil can cause poor atomization thus cannot distribute and mix well with the air inside the engine cylinder resulting in poor combustion and decrease of the diesel engine performance [3].

Several techniques can be adapted to decrease the viscosity of straight vegetable oil such as degumming, transesterification and fuel blending. When the viscosity of the fuel can be improved, the problem related to the viscosity of the straight vegetable oil can be reduced thus increased the diesel engine performance [3]. There are many experimental works conducted by researchers around the world to understand the combustion characteristics of vegetable oil in diesel engine [4–9].

It was proven that a diesel engine can run alternatively on palm oil [2]. However, the high viscosity of palm oil may affect the performance of the diesel engine. Due to high viscosity properties of palm oil, it produced poor atomization process that tends to decrease the performance of the engine. Therefore, in this research optimization on engine performance using palm oil diesel blends will be based on different injection pressure and timings.

3 Experimental Set-up and Procedures

The experiments were conducted by running palm oil and its blends on a diesel engine. The data acquired were recorded onto the table so that analysis can be conducted and the performance curve for palm oil blend can be obtained. The injection timing and pressure were altered to observe the performance of the diesel engine for every configuration. Design of experiment software by Design Expert was used to analyze and optimize the best engine setting.

3.1 The Engine

In this experiment, a single cylinder diesel engine (Yanmar 178F compression ignition) was mounted on the test rig. The engine is mounted with a dynamometer using a belting system. The engine power was rated at 4 kW at the maximum speed of 3600 rpm. The engine specification and details can be seen in Table 1. The experimental setup for this work is shown in Fig. 1.

3.2 Palm Oil Diesel Blends

Palm oil is blended with diesel with specific quantities and particular palm oil palm oil diesel blends were produced. Five different blends of palm oil were used in this

Table 1 Specification of the compression ignition (CI) engine

Model name	Yanmar 178F
Type	Single cylinder, vertical, 4 stroke, air-cooled, direct injection
Bore × stroke	7.8 × 6.2 cm
Displacement	296 cm^3
Compression ratio	20:1
Rated power/speed	4 kW at 3600 rpm
Rotation direction (view from wheel)	Clockwise
Nozzle pressure	200 kgf/cm^2
Fuel tank capacity	3.5 L

Fig. 1 Yanmar 178F CI engine performance experimental setup

Table 2 Five difference mixture of palm oil diesel blends and their density

Fuel blends	Composition	Density (g/ml)
Diesel	100% diesel	0.828
20% PO	20% palm oil + 80% diesel	0.8386
40% PO	40% palm oil + 60% diesel	0.8632
60% PO	60% palm oil + 40% diesel	0.8784
80% PO	80% palm oil + 20% diesel	0.8896
100% PO	100% palm oil	0.9046

Table 3 Fixed and variable parameters in the experiment

Fixed parameters	Variable parameters
Fuel blends ratio	20, 40, 60, 80 and 100% PO
Engine RPM	1000, 1200, 1400, 1600 and 1800 rpm
Engine load	200, 400, 600, 800 and 1000 W
Engine performance (output)	Torque, brake power, BSFC, Bmep, ηth
Injection pressure	200 (original setting), 250 and 300 bar
Injection timing	11°, 14° (original setting) and 17° BTDC (1, 2 and 3 copper shims)

experiment and the details are shown in Table 2. The details of the fixed and variable parameters as well as fuel blends can be seen in Table 3.

The five palm-oil blends were used to run a diesel engine to gather the performance. The performance of the engine such as torque, brake power and bsfc was compared to the diesel fueled engine. For optimization, sets of experiment were obtained using different configurations of injection pressure, injection timing and palm oil blends with the arrangement assisted by DOE (central composite design). Then, the actual data was obtained from the experiments for every configuration. The experimental design table is plot using the actual performance data obtained to acquire the theoretical projection of the engine setting for the best performance of palm oil fueled engine.

3.3 Performance Measurements

An engine dynamometer was connected to the engine shaft of the Yanmar 178F to measure the performance of the diesel engine. A speed sensor was used to measure the speed of the engine in rpm. The dynamometer were attached with a load cell to measure the torque of the engine. While a burette was used to measure the fuel consumption of the engine. To analyze the performance of the engine, the formulae used are as follows.

$$Brake\ Power, BP = 2\pi NT \tag{1}$$

where N is the engine speed in rps and T is the torque produced by the engine. The brake specific fuel consumption can be calculated using Eq. (2);

$$BSFC = \frac{\dot{m}_f}{BP} \tag{2}$$

where $\dot{m}_f$ is the mass flow rate of the fuel, while BP is the brake power produce by the engine.

3.4 Optimization

Optimization of the diesel engine was conducted using a design of experiment software (central composite design). The data is key in the software with the lowest, mean and highest value. The suggestion from the software will be generated for the injection pressure and timing. A test was then conducted using the suggestion from the software.

4 Results and Discussion

4.1 Engine Performance at the Original Setting

The average of torque and brake power comparison to diesel at an engine load range from 200 to 1000 W for palm oil diesel blends are shown in Table 4. In this work, the percentage comparison for torque and brake power to diesel has the same value. The analysis shows that palm oil diesel blends produced a better engine performance when compared to diesel at low speed (below 1400 rpm), except for P80 at 1200 rpm and P100 at 1000 rpm. However, at higher speed (beyond 1400 rpm), all blends engine performance had decreased and were worst when compared to diesel fueled engine.

Table 5 shows the comparison of palm oil diesel blends of average brake specific fuel consumption to diesel at engine load range from 200 to 1000 W. Lower value of bsfc are desired since it indicates that a low amount of fuel is used to produced the power. For all blends P100 had shown the highest bsfc and it has

Table 4 Average torque and brake power of palm oil diesel blends comparison to diesel using original injection setting

Speed (rpm)	Blends ratio (%)				
	P20	P40	P60	P80	P100
1000	1.56	2.44	1.27	13.76	−7.80
1200	6.14	1.17	1.32	−1.54	0.44
1400	2.15	2.36	1.18	4.24	2.09
1600	−1.78	−3.22	−3.46	−0.67	−7.74
1800	−5.45	−9.93	−1.48	−7.22	−3.30

Table 5 Average brake specific fuel consumption of palm oil diesel blends comparison to diesel at the original injection setting

Speed (rpm)	Blends ratio (%)				
	P20	P40	P60	P80	P100
1000	−4.78	3.47	0.60	−18.01	45.46
1200	−11.39	−0.23	2.89	8.13	19.14
1400	−8.83	−7.12	−4.11	−0.30	11.70
1600	11.07	19.46	32.58	27.91	42.44
1800	29.56	43.26	19.56	31.59	27.68

45.56% bsfc when compared to diesel (at 1000 rpm). For other blends (P20–P80) they have a lower bsfc when compared to diesel except for certain conditions (indicated by positive value of bsfc). The best bsfc achieved is 18.01% lower than diesel by P80 at 1000 rpm. Beyond 1400 rpm, all palm oil blends are worse when compared to the diesel fuelled engine.

The average brake mean effective pressure of palm oil blends comparison to diesel at engine load range from 200 to 1000 W are shown in Table 6. At the range of 1000–1400 rpm, all palm oil diesel blend had shown better bmep when compared to diesel except for P100 at 1000 rpm and P80 at 1200 rpm. However, beyond 1400 rpm all blends had shown a low value of bmep when compared to the diesel fuelled engine. The poorest bmep is 9.93% lower than diesel fuel at 800 rpm by P40. The best bmep was produced by P80 which is about 13.76% higher than diesel at 1000 rpm.

The overall performance of the engine is shown by the value of brake thermal efficiency. Table 7 shows the average brake thermal efficiency of palm oil diesel blends comparison to diesel at engine load range from 200 to 1000 W. Once again, at low speed (1000–1400 rpm) all blends had shown better engine performance when compared to diesel except for P100 at 1000 rpm. The thermal efficiency achieved by the P60 at about 21.91% higher than diesel fuelled engine at 1400 rpm. At higher engine speed (1600–1800 rpm) all palm oil diesel blends produced lower value of thermal efficiency when compared with a 14.83% lower value when

Table 6 Average brake mean effective pressure of palm oil diesel blends comparison to diesel at the original injection setting

Speed (rpm)	Blends ratio (%)				
	P20	P40	P60	P80	P100
1000	1.56	2.44	1.27	13.76	−7.80
1200	6.14	1.17	1.32	−1.54	0.44
1400	2.45	2.69	1.34	4.83	2.38
1600	−1.78	−3.22	−3.46	−0.67	−7.74
1800	−5.45	−9.93	−1.48	−7.22	−3.30

Table 7 Average brake thermal efficiency of palm oil diesel blends comparison to diesel at the original injection setting

Speed (rpm)	Blends ratio (%)				
	P20	P40	P60	P80	P100
1000	6.15	6.25	8.43	17.28	−13.35
1200	13.51	5.70	2.01	4.51	1.26
1400	17.62	19.88	21.91	16.15	10.38
1600	−6.02	−11.25	−12.16	−7.92	−14.83
1800	−11.91	−14.65	−9.30	−11.65	−10.74

compared to diesel. The lowest value of thermal efficiency was produced by P100 with the value is at about 14.83% lower when compared to the diesel fuelled engine at 1600 rpm.

Based on the data of the original injection setting, it was found that the palm oil diesel blends can match or exceed the diesel fueled engine performance at low speed in between 1000–1400 rpm. However at higher engine speed, which is beyond 1600 rpm, the engine performance starts to decrease mostly due to the fact that its higher viscosity contributes to a longer combustion period. However at higher speed, the time available for the combustion process has become shorter. Therefore a modification on the injection process is required to improve the mixing quality of fuel and air thus to improve the combustion process.

4.2 Optimization of Palm Oil Diesel Blends Using DOE

The optimization of palm oil diesel blends combustion process was performed at 1800 rpm of engine speed by using a design of experiment software to estimate the number of trials based on different configuration of injection pressure and timing. The injection pressure was varied from 200 to 300 bar and the injection timing from 11° to 17° before top dead center (BTDC). In this work, the injection timing was varied by the number of injector pump shims. 3 pieces of shims will produced 17° btdc injection timing, 2 pieces of shims produced 14° btdc of injection timing (original setting) and 1 piece of shim produced 11° btdc of injection timing. The aim of optimization is to increase the engine performance in terms of torque, brake power and bsfc. Figure 2 shows the suggested configuration for the optimization study.

From the actual data obtained, the software conducted a detailed analysis and suggested the theoretical projection of the engine setting for the best performance of the palm oil fueled engine. The constrained goals were set so that the suggestion was in the range with the experimental factor. Figure 3 shows the theoretical suggestion for the engine configuration for best performance of the engine.

Design Expert software analysis suggested about 12 configurations and ranked them from the most desired configuration to less. The analysis was done on behalf of the actual data that has been collected during the experiment. The desirability of the configuration and performance were ranked based on the theoretical performance of the engine. It can be said that, with the use of a low percentage of palm oil, the higher injection pressure was desired so that the atomization of the fuel was better. Due to better and the finer fuel droplet produced, the shorter time was needed to combust the fuel completely, thus the retard injection timing was expected.

The analysis was only catered to maximize the performance on engine torque and BP only, regardless of the fuel consumption of the engine. Fuel droplets were present at the exhaust manifold during the retard injection timing experiments

Std	Run	Block	Factor 1 A:Injection Time Pieces	Factor 2 B:Injection Pressure kgf/cm2	Factor 3 C:Palm oil blend %PO	Response 1 Torque N.m	Response 2 Brake Power kW	Response 3 BSFC g/kW
1	1	Block 1	1.00	200.00	20.00			
9	2	Block 1	1.00	250.00	60.00			
8	3	Block 1	3.00	300.00	100.00			
3	4	Block 1	1.00	300.00	20.00			
15	5	Block 1	2.00	250.00	60.00			
16	6	Block 1	2.00	250.00	60.00			
17	7	Block 1	2.00	250.00	60.00			
19	8	Block 1	2.00	250.00	60.00			
10	9	Block 1	3.00	250.00	60.00			
11	10	Block 1	2.00	200.00	60.00			
14	11	Block 1	2.00	250.00	100.00			
2	12	Block 1	3.00	200.00	20.00			
4	13	Block 1	3.00	300.00	20.00			
18	14	Block 1	2.00	250.00	60.00			
5	15	Block 1	1.00	200.00	100.00			
20	16	Block 1	2.00	250.00	60.00			
7	17	Block 1	1.00	300.00	100.00			
6	18	Block 1	3.00	200.00	100.00			
13	19	Block 1	2.00	250.00	20.00			
12	20	Block 1	2.00	300.00	60.00			

Fig. 2 Suggested experimental configuration for palm oil blends optimization on torque, brake power and bsfc

Constraints

Name	Goal	Lower Limit	Upper Limit	Lower Weight	Upper Weight	Importance
Injection Timing	is in range	1	3	1	1	3
Injection Pressure	is in range	200	300	1	1	3
Palm oil blend	is in range	20	100	1	1	3
Torque	maximize	6.63	8.79	1	1	3
Brake Power	maximize	1.15362	1.65688	1	1	3

Solutions

Number	Injection Timing	Injection Press	Palm oil blend	Torque	Brake Power	Desirability	
1	3.00	287.57	20.00	8.76116	1.5756	0.910	Selected
2	3.00	286.71	20.00	8.77003	1.57381	0.910	
3	3.00	285.23	20.00	8.78429	1.57073	0.909	
4	3.00	298.08	20.00	8.6185	1.59743	0.901	
5	3.00	274.79	20.00	8.84914	1.54904	0.886	
6	3.00	246.60	20.00	8.71163	1.4905	0.803	
7	2.25	299.90	20.00	8.35486	1.51546	0.758	
8	1.00	229.70	100.00	8.17951	1.48388	0.686	
9	1.00	229.23	100.00	8.17692	1.48442	0.686	
10	1.01	229.69	100.00	8.18132	1.48334	0.686	
11	1.36	242.08	100.00	8.24753	1.45114	0.665	
12	1.78	222.74	100.00	8.12001	1.448	0.635	

Fig. 3 Design goals and theoretical suggestion for best engine configuration by the DOE software

Table 8 Comparison between optimized configuration and original injection setting at 1800 rpm

Solution no.	Injection timing (degree BTDC)	Injection pressure (bar)	Blend ratio (%)	Optimized brake torque (Nm)	Initial brake torque (Nm)	% Of improvement
1	17	288	20	8.8	7.8	12.82
2	17	287	20	8.8	7.8	12.82
3	17	285	20	8.8	7.8	12.82
4	17	298	20	8.6	7.8	10.26
5	17	275	20	8.8	7.8	12.82
6	17	247	20	8.7	7.8	11.54
7	15	300	20	8.4	7.8	7.69
8	11	230	100	8.2	7.9	3.80
9	11	229	100	8.2	7.9	3.80
10	11	230	100	8.2	7.9	3.80
11	12	242	100	8.2	7.9	3.80
12	13	223	100	8.1	7.9	2.53

(3 pieces of shims is equal to 11° btdc). From the presence of the fuel droplets, it can be that the fuel did not burn completely inside the engine cylinder, thus the residue was ejected out during the exhaust stroke. The comparison between the optimized configuration and the original injection setting at 1800 rpm is shown in Table 8.

5 Conclusion

The experimental setup were configured well and the experiment was conducted successfully without any major problem during the experimental run. The modified engine run well despite the modification on the injection timing, injection pressure and usage of difference palm oil blends. The usage of Design Expert software was able to predict and list the possible setting for the diesel engine for maximum performance on palm oil fueled engine (see Table 8). Increase in injection pressure and retard the injection timing would increase the performance of P20 at about 12.82% for engine torque. The objectives of these experiments were achieved.

Acknowledgements The authors would like to thank Universiti Kuala Lumpur Malaysian Spanish Institute for the financial support for this research work via UniKL STRG grant UniKL/CoRI/strl5028.

References

1. Altın, R., Çetinkaya, S., Yücesu, H.S.: The potential of using vegetable oil fuels as fuel for diesel engines. Energ. Convers. Manag. **42**, 529–538 (2001)
2. Almeida, S.C.A.D., Belchior, C.R., Nascimento, M.V.G., et al.: Performance of a diesel generator fueled with palm oil. Fuel **81**, 2097–2102 (2004)
3. Agarwal, D., Kumar, L., Agarwal, A.K.: Performance evaluation of a vegetable oil fueled compression ignition engine. Renew. Energ. **33**, 1147–1156 (2008)
4. Dermibas, A.: Biodiesel from vegetable oils via transesterification in supercritical methanol. Energ. Convers. Manag. **43**, 2349–2356 (2002)
5. Wang, Y.D., Al-Shemmeri, T., Eames, P., et al.: An experimental investigation of the performance and gaseous exhaust emissions of a diesel engine using blends of vegetable oil. Appl. Therm. Eng. **26**, 1684–1691 (2006)
6. Benjumea, P., Agudelo, J., Agudelo, A.: Basic properties of palm oil biodiesel-diesel blends. Fuel **87**, 2069–2075 (2008)
7. Haldar, S.K., Ghosh, B.B., Nag, A.: Studies on the comparison of performance and emission characteristics of a diesel engine using three degummed non-edible vegetable oils. Biomass Bioenerg. **33**, 1013–1018 (2009)
8. Bari, S., Lim, T.H., Yu, C.W.: Effect of preheating of crude palm oil (CPO) on injection system, performance and emission of a diesel engine. Renew. Energ. **27**, 339–351 (2002)
9. Lin, B.F., Huang, J.H., Huang, D.Y.: Experimental study of the effect of vegetable oil methyl ester on diesel engine performance characteristics and pollutant emissions. Fuel **88**, 1779–1785 (2009)

The Potential of Improving the Mg-Alloy Surface Quality Using Powder Mixed EDM

M. A. Razak, A. M. Abdul-Rani, A. A. Aliyu,
Muhamad Husaini Abu Bakar, M. R. Ibrahim, J. A. Shukor,
A. Abdullah, M. Rezal, M. F. Haniff and F. Saad

Abstract The conventional electro-discharge machining (C-EDM) method with low material removal rate (MRR) and high electrode wear rate (EWR) results in high production costs. The C-EDM also suffers from inconsistent machined surface quality and the formation of micro cracks and craters on the machined surface. The objective of this article is to review the advantages of the powder mixed EDM (PMEDM) method and its potential to improve the Mg-alloy surface quality. Research articles related to PMEDM process since the year 2015 until 2018 are summarized in this article. The addition of conductive particles in the dielectric fluid leads to an increase of the spark gap size, which subsequently results in a reduction in electrical discharge power density. The melted and deposited zinc particles on the Mg-alloy machined surface will modify the surface.

M. A. Razak (✉) · M. H. Abu Bakar · M. R. Ibrahim · J. A. Shukor · A. Abdullah
Manufacturing Section, Universiti Kuala Lumpur Malaysian Spanish Institute, Kulim
Hi-Tech Park, 09000 Kulim, Kedah, Malaysia
e-mail: alhapis@unikl.edu.my

M. H. Abu Bakar
e-mail: muhamadhusaini@unikl.edu.my

M. R. Ibrahim
e-mail: mohdriduan@unikl.edu.my

J. A. Shukor
e-mail: azulhisham@unikl.edu.my

A. Abdullah
e-mail: aznizam@unikl.edu.my

A. M. Abdul-Rani · A. A. Aliyu
Mechanical Engineering Department, Universiti Teknologi PETRONAS,
32610 Bandar Seri Iskandar, Perak, Malaysia
e-mail: majdi@utp.edu.my

A. A. Aliyu
e-mail: garoabdul@gmail.com

© Springer Nature Switzerland AG 2019
M. H. Abu Bakar et al. (eds.), *Progress in Engineering Technology*,
Advanced Structured Materials 119, https://doi.org/10.1007/978-3-030-28505-0_4

Keywords Mg-alloy · Mixing powder · PMEDM · Surface quality

1 Introduction

The conventional electro-discharge machining (C-EDM) method with low material removal rate (MRR) and high electrode wear rate (EWR) will be resulting in a high production costs. The C-EDM also suffers of inconsistent machined surface quality and the formation of micro cracks and craters on the machined surface. Furthermore, the application of electric current changes the mechanical properties of the machined surface [1]. Since it was first invented, the EDM technology has been through some stages of improvement. In recent years, new exploratory research works have been initiated to improve the efficiency of the EDM process using the powder mixed EDM (PMEDM) method [2, 3]. The objective of this article is to review the advantages of the PMEDM method and its potential to improve the Mg-alloy surface quality.

The PMEDM method may lead to the improvement of the machined part surface finish, increase the MRR and reduce the EWR and also modifying the machined surface characteristics. Since the disadvantages of the C-EDM method in machining biomedical implants are the fatigue performance due to high corrosion rate of the machined surface, micro cracks, brittle oxidized surface recast layer and internal tensile stresses [4], the PMEDM method is believed to have the potential for surface modification on the machined surface properties, topography and elements composition, thus the machined surface corrosion rate will be reduced. The mixing particles in the dielectric fluid help to improve the sparking efficiency during the ignition process.

2 Working Principle of PMEDM

The addition of conductive particles in the dielectric fluid leads to an increase of the spark gap size, which subsequently will be resulted in a reduction in electrical discharge power density. From the literature of PMEDM researches, a newly

M. Rezal · M. F. Haniff
Electrical, Electronics and Automation Section, Universiti Kuala Lumpur Malaysian Spanish Institute, Kulim Hi-Tech Park, 09000 Kulim, Kedah, Malaysia
e-mail: mrezal@unikl.edu.my

M. F. Haniff
e-mail: mohamadfadzli@unikl.edu.my

F. Saad
Mechanical Section, Universiti Kuala Lumpur Malaysian Spanish Institute, Kulim Hi-Tech Park, 09000 Kulim, Kedah, Malaysia
e-mail: fazidah@unikl.edu.my

fabricated operating tank namely PMEDM operating tank is placed in the original operating tank. The reasons are to ensure that the mixing powder in the dielectric fluid will not be filtered by the fine dielectric fluid filter presence in the original machine reservoir, to avoid any possible damage to the original dielectric circulation system and to reduce the quantity of powder used due to the large size of the original operating tank and huge quantity of dielectric fluid. The newly fabricated PMEDM circulation system can be divided into two systems, which are the closed-loop circulation system and the opened-loop circulation system. Figure 1 presents the schematic diagram of the closed-loop circulation system.

The PMEDM circulation system is equipped with a stirrer to ensure that the powder is circulated inside the PMEDM operating tank and not accumulated at the bottom of the tank and a circulating pump with nozzle directed towards the spark gap to ensure the presence of powder at the cutting area. The disadvantage of the closed-loop circulating system is that the same dielectric fluid and debris would be compounded in PMEDM operating tank throughout the process. The opened-loop circulation system as presented in Fig. 2 has additional mechanisms such as PMEDM reservoir with stirrers, dielectric fluid supply pump, hoses and filter [5–13]. The opened-loop circulation system is used in this research because the PMEDM operating tank is connected to an external reservoir to circulate the renewal dielectric fluid during the process and the 1 µm filter will filter the debris from the Mg-alloy workpiece and copper electrode and allows the zinc powder with an average size of 80 nm to pass through.

In the PMEDM method, the effects of mixing powder on the output response depend mainly on its physical properties, particle size and powder concentration. The main characteristics to consider in the selection of the powder are the electrical conductivity, thermal conductivity, density, melting point, powder size and powder

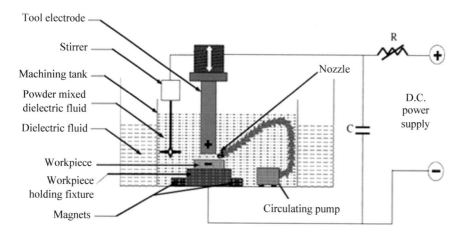

Fig. 1 Diagram of the PMEDM closed-loop circulation system. Adapted from [5]

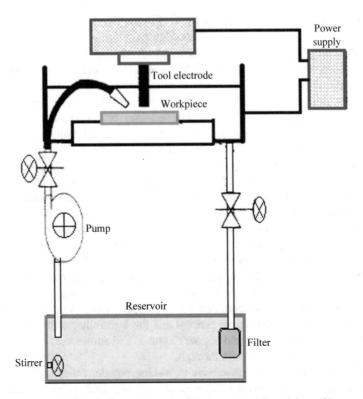

Fig. 2 Diagram of the PMEDM open-loop circulation system. Adapted from [8]

concentration [8]. A metal powder with high electrical conductivity will easily promote more electrons to ionize or breakdown the dielectric fluid for quicker creation of electrical discharge which flows widely in the cutting area. High thermal conductivity of the powder will easily dissipate heat from the cutting area to the dielectric fluid. The low density of the powder will contribute to the suspension of powder in the dielectric fluid. The lowest melting point of the powder will contribute to melting the powder presence in the cutting zone and get deposited on the machined surface to improve its characteristics. The fine powder improves machining rates and surface roughness. The metallic powder grain size can be classified as very coarse (>1000 μm), coarse (355–1000 μm), moderately fine (180–355 μm), fine (125–180 μm) and very fine (90–125 μm) [14]. The amount of powder in the dielectric fluid will promote more electrons to be energized, contributing to faster breakdown of the dielectric fluid and evenly distribution of the electric discharge.

3   Research Articles on PMEDM

Recently, many research articles regarding PMEDM can be found in studies of the different workpiece and mixing powder materials. The advantages of the PMEDM method addressed in those research articles include improved machining efficiency in terms of machining time, MRR, EWR, reducing surface roughness and potential for surface modification. The discussions on surface modification include powder deposition on the machined surface, surface morphology, recast layer and surface hardness. Table 1 summarizes research articles related to the PMEDM process since the year 2015 until 2018.

Table 1   Research articles on PMEDM

Researcher	Year	Workpiece	Mixing powder	Research findings
Bhattacharya et al. [15]	2015	Die steels (AISI D2, AISI D3 and AISI H13)	Graphite, tungsten and titanium	Confirmed material migration from the suspended powder in the dielectric, reduced EWR and increased micro hardness
Prakash et al. [16]	2015	Titanium alloy	Silicon	Increased micro hardness and optimized surface roughness
Tawfiq and Hameed [17]	2015	Die steel (AISI D3, AISI D6 and AISI H13)	Manganese, aluminum and aluminum-manganese	Minimized surface roughness with manganese powder
Syed et al. [18]	2015	Aluminum metal matrix composite	Aluminum	MRR was improved with PMEDM method
Long et al. [19]	2016	Die steels (SKD 61, SKD 11 and SKT 4)	Titanium	MRR increased by 42.1% compared to C-EDM
Bembde and Sawale [20]	2016	Steel (EN31)	Silicon	Optimized MRR and surface roughness
Long et al. [21]	2016	Mould steels (SKD 61, SKT 4 and SKD 11)	Titanium	Optimized surface roughness, uniform thickness of the white layer, increased hardness and reduced cracks
Arya et al. [22]	2016	Stainless steel (304)	Copper	Increased MRR and decreased EWR
Kumar et al. [23]	2016	Steel (HCHCr)	Silicon	Increased MRR by 41% and decreased EWR by 35%
Prakash et al. [24]	2016	Titanium alloy	Silicon	Reduced recast layer by increased powder concentration, improved MRR and reduced EWR

(continued)

Table 1 (continued)

Researcher	Year	Workpiece	Mixing powder	Research findings
Kumar et al. [25]	2016	Titanium alloy	Titanium, tin, vanadium, manganese and tungsten	Optimized EWR and observed the hard and brittle black layer formed on the electrode surface
Abdul-Rani et al. [26]	2016	Titanium alloy	Aluminum	Reduced surface roughness. Less cracks, craters and voids on machined surface
Abdul-Rani et al. [27]	2017	Titanium alloy	Aluminum	Improved surface morphology and reduced rate of corrosion
Banh et al. [28]	2017	Die steel	Titanium	Concentration of the powder was identified as the most significant input compared to other operation inputs. Tool wear and surface roughness have decreased. Increased accuracy and machining efficiency
Chaudhury et al. [29]	2017	Steel (EN19)	Tungsten	Powder concentration was identified as the most significant input affecting the MRR
Karunakaran and Chandrasekaran [30]	2017	Nickel alloy (Inconel 800)	Aluminum	Increased MRR by 36.67% compared to C-EDM
Bhaumik and Maity [31]	2017	Stainless steel	Silicon carbide	Most significant operation input was peak current. Surface irregularities increased when peak current increased. Surface roughness decreased when powder concentration increased. Cracks, craters and globules were formed on the machined surface
Nanimina et al. [32]	2017	Molybdenum high-speed steel	Aluminum	The added surfactant was not significant. Surface quality was slightly improved
Shabgard and Khosrozadeh [33]	2017	Titanium alloy	Carbon nanotubes	Reduced size of micro cracks, MRR and EWR. Surface roughness decreased as reduction of spark energy and uniform spark distribution
Ou and Wang [34]	2017	Titanium alloy	Hydroxyapatite	Reduced surface roughness and recast layer. MRR was decreased. Calcium and phosphorus were found embedded on the recast layer

(continued)

Table 1 (continued)

Researcher	Year	Workpiece	Mixing powder	Research findings
Prakash et al. [35]	2018	Mg-alloy	Hydroxyapatite	Found hydroxyapatite layer with interconnected pores of 5–10 μm
Kumar et al. [36]	2018	Inconel 825	Aluminum oxide	MRR and surface roughness are directly influenced by peak current, pulse on-time and gap voltage. PMEDM improves surface roughness

4 Improving Mg-Alloy Surface Quality Using PMEDM

Figure 3 illustrates how the suspended powder in the dielectric fluid is resulting in a better machined surface quality. During the PMEDM process, the powder suspends in the dielectric fluid reduces the insulating strength of the dielectric fluid, reduces the energy density on the workpiece and increases the spark gap distance between the workpiece and tool electrode. Thus, the plasma channel will enlarge and widen. With optimum powder concentration, the process becomes more stable with higher spark frequency generated and ensures a homogeneous distribution of the discharge energy which is creating uniform erosion from the workpiece and results in shallow craters that served to improve the surface finish [23, 34, 37]. The improvement of the surface quality is not only on the roughness, but also to reduce the micro cracks, globules and voids.

In order to understand the formation of the bridge in the spark gap between the electrode and Mg-alloy workpiece, assume that the added zinc particles are spheres with higher permittivity than the dielectric fluid. The zinc particles in the gap get polarized in an electric field and experience a force toward the place of maximum

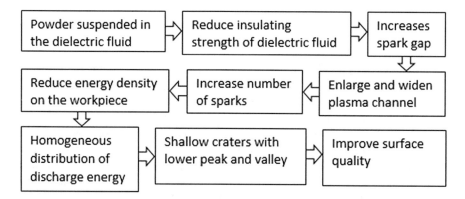

Fig. 3 The process of the influence of powder mixed dielectric fluid in improving surface quality. Adapted from [5, 7, 9, 13, 26, 28, 37–39]

stress, and in a uniform electric field that can usually be developed by a small sphere gap, the field is the strongest in the uniform field region. Thus, the force on the particle is zero and the particle remains in equilibrium. Therefore, the particles will be dragged into the uniform field region. Since the permittivity of the zinc particles is higher than that of the dielectric fluid, the presence of particles in the uniform field region will cause a flux concentration at its surface. Other particles also tend to move towards the higher flux concentration. If the present zinc particles are large, they become aligned due to these forces and form a highly conductive bridge across the spark gap. Hence, this will lead to easier liquid breakdown [40]. However, the concentration of the powder is important. Too high concentration will be resulting in a negative output.

The powder suspended in the dielectric fluid is incorporated into a recast layer on the workpiece to obtain a surface with the desired properties, which is interesting from the point of view of applications involving surface modifications. The machined surface roughness and the recast layer thickness depending on the melting point and thermal conductivity of the material, since the removal mechanism is based on spark erosion. During the spark erosion process, more material will be melted and evaporated by electric discharge when the workpiece has a lower melting point [34]. Based on this reason, since the Mg-alloy has a considerably lower melting point, it is easier for zinc particles to deposit and bound on the machined surface. The additive particle size is determinative in machined surface quality. The smallest particles between 70 and 80 nm produced the best surface finish while simultaneously increasing the recast layer thickness [41]. The recast layer incorporated with zinc particles will improve the corrosion resistance of Mg-alloy machined surface.

The higher powder concentration leads to a smaller gap distance, which will allow less dielectric fluid passing through the sparking gap and taking away heat from the workpiece. This causes heat accumulation on the workpiece surface, which is beneficial for the formation of the recast layer.

5   Conclusion

From the review of previous research works, it can be concluded that the PMEDM method has the potential to improve the Mg-alloy surface quality. The high temperature from the plasma channel melts the workpiece accompanied by the mixing powder incorporating into the surface. Increasing discharge current provides the high current density of the discharge channel resulting in debris being rapid ejected out of the gap by mechanical impaction. However, raising the pulse on-time promotes the amounts of the powder on the machined surface due to the discharge channel expansion. Since the micro cracks can cause a disruption of the surface oxide layer which expedites the corrosion [42], the melted and deposited zinc particles on the Mg-alloy machined surface will modify the surface. Surface modification of the machined surface is not only beneficial for improving the corrosion resistance. It will also change the surface hardness. The type of mixing powder significantly affects the micro hardness of the machined surface [43].

References

1. Nguyen-Tran, H.-D., Oh, H.-S., Hong, S.-T., Han, H.N., Cao, J., Ahn, S.-H., et al.: A review of electrically-assisted manufacturing. Int. J. Precis. Eng. Manufact. Green Technol. **2**, 365–376 (2015)
2. Razak, M.A., Abdul-Rani, A.M., Nanimina, A.M.: Improving EDM efficiency with silicon carbide powder-mixed dielectric fluid. Int. J. Mater. Mech. Manufact. **3**, 40–43 (2015)
3. Razak, M.A., Abdul-Rani, A.M., Rao, T.V.V.L.N., Pedapati, S.R., Kamal, S.: Electrical discharge machining on biodegradable AZ31 magnesium alloy using Taguchi method. Procedia Eng. **148**, 916–922 (2016)
4. Strasky, J., Janecek, M., Harcuba, P.: Electric discharge machining of Ti-6Al-4V alloy for biomedical use. In: WDS, pp. 127–131 (n.d.)
5. Kansal, H., Singh, S., Kumar, P.: Effect of silicon powder mixed EDM on machining rate of AISI D2 die steel. J. Manufact. Process. **9**, 13–22 (2007)
6. Zhao, W., Meng, Q., Wang, Z.: The application of research on powder mixed EDM in rough machining. J. Mater. Process. Technol. **129**, 30–33 (2002)
7. Kumar, S., Singh, R., Singh, T., Sethi, B.: Surface modification by electrical discharge machining: a review. J. Mater. Process. Technol. **209**, 3675–3687 (2009)
8. Tzeng, Y.F., Lee, C.Y.: Effects of powder characteristics on electrodischarge machining efficiency. Int. J. Adv. Manufact. Technol. **17**, 586–592 (2001)
9. Nanimina, A.M., Rani, A.M.A., Ginta, T.L.: Assessment of powder mixed EDM: a review. In: MATEC Web of Conferences, p. 04018 (2014)
10. Singh, A., Singh, R.: Effect of powder mixed electric discharge machining (PMEDM) on various materials with different powders: a review. Int. J. Innovative Res. Sci. Technol. **2**, 164–169 (2015)
11. Mohal, S., Kumar, H., Kansal, S.: Nano-finishing of materials by powder mixed electric discharge machining (PMEDM): a review. Sci. Adv. Mater. **7**, 2234–2255 (2015)
12. Kansal, H., Singh, S., Kumar, P.: Technology and research developments in powder mixed electric discharge machining (PMEDM). J. Mater. Process. Technol. **184**, 32–41 (2007)
13. Kumar, S., Batra, U.: Surface modification of die steel materials by EDM method using tungsten powder-mixed dielectric. J. Manufact. Process. **14**, 35–40 (2012)
14. Liu, Y., Wang, C., Zhang, Y., Xiao, S., Chen, Y.: Fractal process and particle size distribution in a TiH 2 powder milling system. Powder Technol. **284**, 272–278 (2015)
15. Bhattacharya, A., Batish, A., Bhatt, G.: Material transfer mechanism during magnetic field–assisted electric discharge machining of AISI D2, D3 and H13 die steel. Proc. Inst. Mech. Eng. Part B J. Eng. Manufact. **229**, 62–74 (2015)
16. Prakash, C., Kansal, H., Pabla, B., Puri, S.: To optimize the surface roughness and microhardness of β-Ti alloy in PMEDM process using non-dominated sorting genetic algorithm-II. In: 2015 2nd International Conference on Recent Advances in Engineering & Computational Sciences (RAECS), pp. 1–6 (2015)
17. Tawfiq, M.A., Hameed, A.S.: Effect of powder concentration in PMEDM on surface roughness for different die steel types. Int. J. Curr. Eng. Technol. **5**, 3323–3329 (2015)
18. Syed, K.H., Anuraag, G., Hemanth, G., Subahan, S.A.: Powder-mixed EDM machining of aluminium-silicon carbide composites. Indian J. Sci. Technol. **8**, 133–137 (2015)

19. Long, B.T., Phan, N.H., Cuong, N., Jatti, V.S.: Optimization of PMEDM process parameter for maximizing material removal rate by Taguchi's method. Int. J. Adv. Manufact. Technol **87**, 1–11 (2016)
20. Bembde, M.S., Sawale, J.K.: Optimization of process parameters of powder mixed dielectric EDM for MRR and Ra by Grey relational analysis method. Int. J. Sci. Res. **5**, 389–391 (2016)
21. Long, B.T., Phan, N.H., Cuong, N., Toan, N.D.: Surface quality analysis of die steels in powder-mixed electrical discharge machining using titan powder in fine machining. Adv. Mech. Eng. **8**, 1–13 (2016)
22. Arya, G., Garg, S., Kumar, E.A.: Experimental investigation to identify the parametric effect on material removal rate and electrode wear rate during PMEDM. Int. J. Eng. Technol. Sci. Res. **3**, 1–5 (2016)
23. Kumar, N., Ale, P., Kumar, N., Sharma, S.: Study of PMEDM efficiency on HCHCr steel using silicon powder in dielectric fluid. Int. J. Emerg. Technol. Eng. Res. **4**, 54–61 (2016)
24. Prakash, C., Kansal, H., Pabla, B., Puri, S.: Experimental investigations in powder mixed electric discharge machining of Ti–35Nb–7Ta–5Zrβ-titanium alloy. Materials and Manufacturing Processes **32**, 1–13 (2016)
25. Kumar, S., Singh, R., Batish, A., Singh, T.: Study the effect of black layer on electrode wear ratio in powder mixed electric discharge machining of titanium alloys. Int. J. Mach. Mach. Mater. **18**, 18–35 (2016)
26. Abdul-Rani, A., Nanimina, A., Ginta, T., Razak, M.: Machined surface quality in nano aluminum mixed electrical discharge machining. Procedia Manufact. **7**, 510–517 (2016)
27. Abdul-Rani A.M., Nanimina A.M., Ginta, T.L.: Surface morphology and corrosion behavior in nano PMEDM. In: Key Engineering Materials, 2017, pp. 61–65 (n.d.)
28. Banh, T.-L., Nguyen, H.-P., Ngo, C., Nguyen, D.-T.: Characteristics optimization of powder mixed electric discharge machining using titanium powder for die steel materials. Proc. Inst. Mech. Eng. Part E J. Process Mech. Eng. p. 0954408917693661 (2017)
29. Chaudhury, P., Samantaray, S., Sahu, S.: Optimization of process parameters of powder additive-mixed electrical discharge machining. In: Innovative Design and Development Practices in Aerospace and Automotive Engineering, ed: Springer, pp. 415–425 (2017)
30. Karunakaran, K., Chandrasekaran, M.: Influence of process parameters in n-PMEDM of Inconel 800 with electrode and coated electrodes. In: MATEC Web of Conferences, p. 02002 (2017)
31. Bhaumik, M., Maity, K.: Effect of machining parameter on the surface roughness of AISI 304 in silicon carbide powder mixed EDM. Decision Sci. Lett. **6**, 261–268 (2017)
32. Nanimina, A., Abdul-Rani, A., Soultan, M., Khayal, M., Krishnan, S.: Surfactant and nano aluminum PMEDM on molybdenum high speed steel. Revue Scientifique du TCHAD **1**, 73–79 (2017)
33. Shabgard, M., Khosrozadeh, B.: Investigation of carbon nanotube added dielectric on the surface characteristics and machining performance of Ti–6Al–4V alloy in EDM process. J. Manufact. Process. **25**, 212–219 (2017)
34. Ou, S.-F., Wang, C.-Y.: Effects of bioceramic particles in dielectric of powder-mixed electrical discharge machining on machining and surface characteristics of titanium alloys. J. Mater. Process. Technol. **245**, 70–79 (2017)
35. Prakash, C., Singh, S., Singh, M., Verma, K., Chaudhary, B., Singh, S.: Multi-objective particle swarm optimization of EDM parameters to deposit HA-coating on biodegradable Mg-alloy. Vacuum **158**, 180–190 (2018)
36. Kumar, V., Kumar, A., Kumar, S., Singh, N.: Comparative study of powder mixed EDM and conventional EDM using response surface methodology. Mater. Today Proc. **5**, 18089–18094 (2018)
37. Bajaj, R., Tiwari, A.K., Dixit, A.R.: Current trends in electric discharge machining using micro and nano powder materials: a review. Mater. Today Proc. **2**, 3302–3307 (2015)
38. Bhattacharya, A., Batish, A., Kumar, N.: Surface characterization and material migration during surface modification of die steels with silicon, graphite and tungsten powder in EDM process. J. Mech. Sci. Technol. **27**, 133–140 (2013)

39. Han, M.-S., Min, B.-K., Lee, S.J.: Improvement of surface integrity of electro-chemical discharge machining process using powder-mixed electrolyte. J. Mater. Process. Technol. **191**, 224–227 (2007)
40. Marashi, H., Jafarlou, D.M., Sarhan, A.A., Hamdi, M.: State of the art in powder mixed dielectric for EDM applications. Precis. Eng. **46**, 11–33 (2016)
41. Yih-Fong, T., Fu-Chen, C.: Investigation into some surface characteristics of electrical discharge machined SKD-11 using powder-suspension dielectric oil. J. Mater. Process. Technol. **170**, 385–391 (2005)
42. Walter, R., Kannan, M.B., He, Y., Sandham, A.: Effect of surface roughness on the in vitro degradation behaviour of a biodegradable magnesium-based alloy. Appl. Surf. Sci. **279**, 343–348 (2013)
43. Batish, A., Bhattacharya, A.: Mechanism of material deposition from powder, electrode and dielectric for surface modification of H11 and H13 die steels in EDM process. Mater. Sci. Forum **701**, 61–75 (2012)

Validation of Driver's Cognitive Load on Driving Performance Using Spectral Estimation Based on EEG Frequency Spectrum

Firdaus Mohamed, Pranesh Krishnan and Sazali Yaacob

Abstract Driver's drowsiness becomes a prominent factor that causes the growing number of a road accident in the past few years and turns out to be perturbing for road safety. This research presents approaches for drowsiness and alertness recognition based on the electroencephalography (EEG) and power spectrum to evaluate the driver's vigilance level in a static driving simulator. The EEG databases are validated using the Karolinska sleepiness scale (KSS) and reaction time (RT). Frequency-domain power spectral density (PSD) feature extraction techniques were evaluated (periodogram, Lomb-Scargle, Thompson multitaper, and Welch) with supervised learning classifiers (MLNN, QSVM, and KNN). The highest accuracy is attained from MLNN using Lomb-Scargle PSD with 96.3% and the minimum accuracy is attained from QSVM and KNN with both 62.2% using periodogram and Welch PSD features set respectively.

Keywords Drowsiness · EEG · KSS · RT

1   Introduction

Severe sleep deficiency, continuously driving the vehicle for extended hours, unbalanced driving schedule were common among the car drivers especially for heavy vehicle drivers. Driving a vehicle under the influences of fatigue and drowsiness will cause longer response time, alertness reduction and deficits in information processing and communication, which may contribute to the increase

F. Mohamed (✉) · P. Krishnan · S. Yaacob
Intelligent Automotive Systems Research Cluster, Universiti Kuala Lumpur
Malaysian Spanish Institute, Kulim Hi-Tech Park, 09000 Kulim, Kedah, Malaysia
e-mail: firdausmohamed86@gmail.com

P. Krishnan
e-mail: praneshkrishnan@gmail.com

S. Yaacob
e-mail: sazali.yaacob@unikl.edu.my

© Springer Nature Switzerland AG 2019
M. H. Abu Bakar et al. (eds.), *Progress in Engineering Technology*,
Advanced Structured Materials 119, https://doi.org/10.1007/978-3-030-28505-0_5

of the accident and lacks accuracy in driving maneuver, especially at high speeds driving. The current technologies from original equipment manufacturer (OEM) for recognizing the driver's fatigue/drowsiness level are still in its early stages and the information of understanding government strategies in preventing road accidents and vehicle manufacturer's approaches are thus far inadequate to avert from deadly road catastrophes.

In recent years, a variety of techniques and methods have been introduced by researchers for recognizing driver's drowsiness based on face recognition, body movements, and physiological signals. Amongst the biosignals based methods as an indicative measure, perhaps the electroencephalography (EEG) signals are being the utmost auspicious measures of driver's drowsiness. Nevertheless, there are some obstacles in designing EEG-based driver drowsiness recognition systems, which comprises, lack of a significant index for detecting fatigue/drowsiness and pervasive noise interferences while recording the EEG activities while driving in a static driving simulator.

Moreover, driving factors such as traffic congestion, driving posture and prolong contact with vehicle noise and vibration may also contribute to driver fatigue/ drowsiness. At the present time, there are no models to differentiate the relationship between the cognitive and physical significances of fatigue that correlate to the driver's vigilance level. Henceforth, it is crucial to design the driver fatigue/ drowsiness recognition model in view of the behavioral characteristics of the driver and environmental aspects to estimate the level of vigilance. The event-related potentials (ERP) observed from the drivers can be used to identify the cognitive state and fatigue index that may improve the driver's performance capacity and prevents a catastrophic incident. However, there is limited information on the correlation between driver fatigues due to driving factors and attitude/behavior. Therefore, understanding the psychology of fatigue may lead to better fatigue-alertness model.

2   Literature Review

Driver drowsiness becomes a major factor that causes road accidents in Malaysia [1–4] for the past few years. The main factor that causes human faults are fatigue and drowsiness that affect the driver's behavior due to sleep deprivation, drinking of alcohol, long driving hours and driving schedules for instance driving at night-time, early dawn, mid-afternoon and particularly in the monotonous traffic congestion, personality and personality may also stimulate fatigue [5, 6]. Thus, preventing such disastrous accidents is thus a major focus of government strategies, vehicle manufacturers and research efforts in the field of automotive and safety research [2, 7].

For the past years, a number of methods and techniques have been proposed to identify vigilance changes such as physical changes during fatigue and measuring physiological changes of drivers, such as eye activity measurement, heartbeat rate, skin electric potential, and especially, brain wave activities as a means of detecting the cognitive states [8–10]. Consequently, analyzing the EEG signals during driver fatigue may be a promising indicator for use in a driver's drowsiness recognition systems. The EEG-based classification of alertness levels has the benefits of producing an accurate and quantitative assessment and a relatively shorter one to track second-to-second fluctuations in the driver's driving performance. Nevertheless, monitoring driver's fatigue based on EEG signals and estimating the level of fatigue index are still in its early stages and there is lot to discover for instance the EEG frequency bands that are associated with fatigue and drowsiness. In addition, to estimate to which degree these cognitive states associated with EEG changes can be effectively integrated into a driver's drowsiness counter-measure system [11–13]. Likewise, there are few obstacles in analyzing the raw EEG signals, which comprises omnipresent noise while acquiring the EEG signals in a static driving simulator [13].

Consequently, in this work, it is recommended to design a driver's vigilance recognition model based on the EEG signals by integrating signal processing techniques and supervised machine learning algorithms to estimate the driver cognitive state while driving a vehicle in a virtual-based static simulator under a monotonous driving environment. To minimize the computational time, the features used for modeling should be minimal [14]. Thus, in this research, it is proposed to minimize the number of features using soft computing techniques and classification using non-linear supervised classification algorithms [15]. The proposed detection model recognizes the distinction between the driver's alertness level whether the driver is in an alert state or in fatigue state due to induced-fatigue driving or cognitive-behavior using the EEG signals. Then the level of alertness is related with sleepiness.

Furthermore, the adaptive model can be utilized to alert drivers and regulators in optimizing the properties of the interface systems in identifying potential catastrophes. The proposed system alerts the driver during fatigue/drowsiness according to the recognition of the cognitive state and produces the fatigue index and level of alertness. The proposed system also helps the driver to be more attentive and intuitive to prevent fatal road accidents.

3　Methodology

3.1　Experimental Setup for Data Recording

This research work involves acquiring EEG data from a number of subjects with a specifically designed protocol for alert driving (AD) and fatigue driving (FD) simulations. In order to acquire the drivers' EEG database, two specifics driving simulation procedures were developed. The first procedure is an alert driving session which is recorded in the morning for AD database and the second procedure is fatigue/drowsiness driving session which is recorded in the afternoon for FD database. The developed EEG databases for AD and FD were collected from 10 normal subjects in two different driving procedures. The proposed methodologies involve the recognition of the driver's alertness level (alert and fatigue) through exclusive mode (subject wise analysis), as illustrated in Fig. 1.

3.2 Driving Tasks

Open source driving simulation called OpenDS [17] is used for the driving simulation tasks. The ReactionTest driving scheme from OpenDS consists of two driving tasks that each subject requires to perform, which are "Slow Down" and "Change Lane", and is displayed in Fig. 2. The subjects were required to respond quickly on the appearing signs. A total of 10 tasks for "Slow Down" and 10 tasks for "Change Lane" the subjects need to perform throughout each of the driving tasks. The subjects were asked to repeat the driving sessions ten times.

3.3 Data Validation

Data validation is commonly performed when developing EEG database to discover whether the acquired data from nominally different tasks are statistically different. There are various methods available for data validation, however, simple and coherent methods of validating data are considered desirable. In this research work, two EEG databases (AD database and FD database) were developed using 10 normal subjects with each subject database comprising of 20 EEG signals (2 driving tasks × 10 trials) and validated using the Karolinska sleepiness scale (KSS).

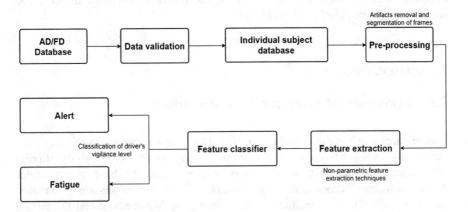

Fig. 1 Block diagram for fatigue and alertness detection model

Fig. 2 Sign (red-X) shows on the display designated for braking and sign (green-arrow) appears on the display designated for lane changing

The KSS is a subjective nine-point scale; ranging from extremely alert (1-point), very alert (2-point), alert (3-point), rather alert (4-point), neither alert nor sleepy (5-point), some sign of sleepiness (6-point), sleepy but no effort to keep awake (7-point), sleepy and some effort to keep awake (8-point), and very sleepy with great effort to keep awake (9-point). The scale was rated after each driving task was completed with the time taken for each completion of the driving task is approximately 4.5 min. This rating scale will determine the significance of the two databases i.e. AD and FD.

From Table 1, it is observed that subject S7 has a minimum KSS value of 2.5 and subject S1 has the maximum KSS value of 6.7. Also, it is discovered that only subject S6 has an alert-KSS value of 3.4 higher than the fatigue-KSS value of 3.1. These values indicated that the data has a significant difference between the developed EEG databases.

3.4 Pre-processing

The acquired brain signals are subjected to pre-processing to eliminate artifacts, segmentation of frames, and frequency bands segmentation. The frame signals are band-pass filtered within 0.1 to 100 Hz (frequency range) and split into six frequency band signals, i.e. delta (δ) 0.1–4 Hz, theta (θ) 4–8 Hz, alpha (α) 8–16 Hz, beta (β) 16–32 Hz, gamma1 ($\gamma 1$) 32–64 Hz, and gamma2 ($\gamma 2$) 64–100 Hz using IIR digital filter. The segmented frequency band signals are used to extract relevant features using four different nonparametric feature extraction methods.

Table 1 Mean KSS rating results for alert and fatigue indexes

Subject (S)	Alert index	Fatigue index
S1	2.8	**6.7**
S2	4.3	6.4
S3	4.1	5.9
S4	4.1	5.6
S5	2.6	3.3
S6	3.4	3.1
S7	**2.5**	5.7
S8	4.6	6.3
S9	5.4	5.7
S10	2.6	3.9

3.5 Feature Extraction

The pre-processed frequency band signals reside valuable information in classifying the driver's vigilance (fatigue and alert) model. However, this data set is quite huge and consumes high computational power to process and resulting in increases of the computational time. This valuable information can be transformed into a reduced-size of a set of features using a relevant feature extraction algorithm. The transformed features set represents desirable information to visualize, verification, and classification of the model. Therefore, feature extraction is the procedure of recognizing assertive features of the EEG signals and developing minimal dimensional feature vectors and thus resulting in the preserving of information in recognizing the driver's vigilance level.

Henceforth, in this work, the pre-processed EEG signals were evaluated using the non-parametric feature extraction algorithm in the frequency domain. The pre-processed signals were also used to extract spectral estimation features using periodogram PSD, Lomb-Scargle PSD, Thompson's multitaper PSD, and Welch PSD technique. Figure 3 represents an overview of the feature extraction techniques used in this research.

The extracted features are correlated with the corresponding driving tasks (Alert Driving and Fatigue Driving) and consequently used as input to feature classifiers for the classification of Alert Driving/Fatigue Driving tasks. The database of extracted features is then validated using k-fold cross-validation and subjected to a machine learning algorithm for the purpose of classification.

3.6 Feature Classifiers

A feature classifier is an essential part of most classification and pattern recognition algorithms. To develop an adaptive fatigue and alertness model, a classifier model has to be developed to recognize the fatigue and alertness state corresponding to the

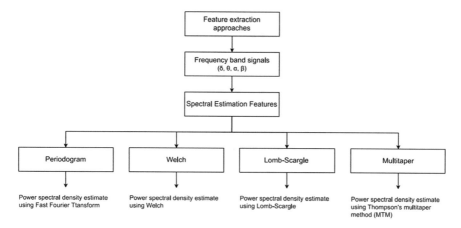

Fig. 3 Overview of the feature extraction approaches used in driver alertness and fatigue recognition (DAFR) model

designed data acquisition protocol. Finding the appropriate classification algorithm is quite challenging because there is no perfect method to accurately estimate the output that fits all. Therefore, selecting the suitable algorithm involves trading off one advantage against another, including accuracy, complexity, and model speed. Higher predictive accuracy can be achieved from the classification learning process by selecting only the relevant characteristics of the data.

Multilayer Neural Network (MLNN)

Training in MLNN involves adjusting the values of the weights and biases of the network to enhance network performance. Through the training procedures, the MLNN adjusts the synaptic weights in response to the input, so that the actual output response is approximately the same as the desired response. In this study, the Levenberg–Marquardt algorithm is used for neural network training. The Levenberg–Marquardt algorithm [16, 17] was developed by Kenneth Levenberg and Donald Marquardt for solving the problem of minimizing a nonlinear function. In neural network training, the Levenberg–Marquardt algorithm combines the features from both gradient descent method and Gauss-Newton method and it is considered as one of the most efficient training algorithms [18] and widely used in backpropagation methods.

Quadratic SVM (QSVM)

Support vector machine (SVM) was developed by Boser et al. [19] in 1992. The SVM can be used for regression and classification of the given input data into two or more classes [20–22]. The SVM performs better than MLNN in terms of training, generalization, no local optimal and scales relatively well to a large scale of dimension [23]. The SVM is a particularly powerful and universal learning machine because of its characteristics such as ease of changing the implemented decision surface and capacity control obtained by optimizing the margin. Basically, SVM works by finding a hyperplane, as in Eq. (1), that separates the positive and negative values of a given set of training vectors (dataset), as in (2), from each other with maximum margin to segregate the output classes, i.e. Alert and Fatigue.

$$\langle w, x \rangle + b = 0 \tag{1}$$

$$\mathcal{D} = \{x_j, y_j\}_{j=1}^{N}, \ x_j \in R^n, \ y_j \in \{-1,1\} \tag{2}$$

If the set of training vectors is separated without miscalculation and the distance between the nearest vector to the hyperplane is maximal, then it is said to be optimally separated by the hyperplane. However, in this work, for a nonlinearly separable data, a quadratic SVM is used to create an optimal hyperplane in a high-dimensional space with maximum margin to perform pattern recognition for a given set of training vectors to segregate the output classes. The general degree-two polynomial kernel is defined as in Eq. (3).

$$K(x, x') = (\langle x, x' \rangle + 1)^2 \tag{3}$$

where x and x' are vectors in the input space, which the polynomial kernel maps a two-dimensional input vectors into a six-dimensional feature space [22].

K-Nearest Neighbor (KNN)

KNN is a non-parametric classifier, supervised learning algorithm, and widely used for pattern classification in EEG studies [24, 25]. In pattern classification, the KNN classifies the feature samples using the nearest training pattern in the feature vectors and the target class is classified based on majority voting of its neighbors. To the extent that, the target class is associated with the most predominant class amongst its KNN measured by a distance function. The KNN classifier significantly depends on the value of k (positive integer) and the distance function used. Numerous methods have been used to optimize the value of k and different distance functions such as Euclidean, Manhattan, Minkowski, and Hamming.

In this study, the KNN algorithm is implemented using the Euclidean distance function to locate the nearest neighbor [26]. The Euclidean distance function measures $ED(x, y)$ between two samples (points) x and y using Eq. (4). The number of neighbor's 'K' is used to classify the new test vector which varies from 1 to 10, to determine the classification accuracy.

$$ED(x, y) = \sqrt{\sum_{i=1}^{k} (x_i - y_i)^2} \tag{4}$$

4   Results and Conclusion

From Table 2, it can be perceived that the performance of MLNN classifier has the maximum classification rate of 96.3% using the Lomb-Scargle PSD features set and a minimum classification rate of 82.0% using the Welch PSD features set.

Table 2 Comparison of maximum classification results

PSD features	Cross-validation	No. of samples	Training samples	Testing samples	No. of trials	Classifiers	Min (%)	SD	Max (%)
Periodogram	10-fold	8640	7776	864	10	MLNN	84.0	5.0	94.0
						QSVM	**62.2**	8.2	78.6
						KNN	68.4	8.5	85.5
Lomb-Scargle	10-fold	8640	7776	864	10	MLNN	85.0	5.6	**96.3**
						QSVM	63.7	7.9	79.5
						KNN	68.2	8.7	85.6
Thompson multitaper	10-fold	8640	7776	864	10	MLNN	83.7	5.2	94.1
						QSVM	63.0	7.7	78.3
						KNN	66.6	10.1	86.7
Welch	10-fold	8640	7776	864	10	MLNN	82.0	6.2	94.3
						QSVM	**62.2**	9.8	81.7
						KNN	**62.2**	9.9	82.1

The performance of QSVM classifier has the maximum classification accuracy of 81.7% using the Welch PSD features set and minimum classification accuracy of 62.2% using the periodogram and Welch PSD features set. The performance of KNN classifier has the maximum classification correctness of 86.7% using the Thompson multitaper PSD features set and minimum classification correctness of 62.2% using the Welch PSD features set.

From the results, it is summarized that the performance of MLNN classifier model has the maximum classification accuracy of 96.3% using the Lomb-Scargle PSD features set and the performance of QSVM and KNN both have the minimum accuracy of 62.2% using the periodogram and Welch PSD features set respectively.

Acknowledgements The authors are grateful to the Ministry of Higher Education (MoHE) Malaysia, for providing the funding of this research (Ref: FRGS/1/2016/TK04/UNIKL/01/1).

References

1. Lal, S.K.L., Craig, A., Boord, P., et al.: Development of an algorithm for an EEG-based driver fatigue countermeasure. J. Safety Res. **34**(3), 321–328 (2003)
2. Mustafa, M.N.: Overview of current road safety situation in Malaysia. First Int. Vis. Inform. Conf. **2015**(1), 26 (2010)
3. Fai, L.C.: Miros statistics say human error causes 80% of traffic accidents. Sun Media Corporation Sdn. Bhd (2015). http://www.thesundaily.my/news/1333889. Accessed 27 Mar 2017
4. Lee, J.: Road deaths in Malaysia above world average—report. In: Cars, Local News (2015). https://paultan.org/2015/10/13/road-deaths-in-malaysia-above-world-average-report/. Accessed 27 Mar 2017
5. Wang, Q., Yang, J., Ren, M., et al.: Driver fatigue detection: a survey. In: World Congress on Intelligent Control and Automation (WCICA), Vol. 2, pp. 8587–8591, Conference Proceedings (2006)
6. Stern, J.A., Boyer, D., Schroeder, D.: Blink rate: a possible measure of fatigue. Hum. Factors **36**(2), 285–297 (1994)
7. Charbonnier, S., Roy, R.N., Bonnet, S., et al.: EEG index for control operators' mental fatigue monitoring using interactions between brain regions. Expert Syst Appl **52**, 91–98 (2016)
8. Liang, S.F., Lin, C.T., Wu, R.C., et al.: Monitoring driver's alertness based on the driving performance estimation and the EEG power spectrum analysis. In: IEEE Engineering in Medicine and Biology Society. Vol. 6, pp. 5738–5741, Conference Proceedings (2005)
9. Lee, B.G., Chung, W.Y.: Driver alertness monitoring using fusion of facial features and bio-signals. IEEE Sens. J. **12**(7), 2416–2422 (2012)
10. Al-Sultan, S., Al-Bayatti, A.H., Zedan, H.: Context-aware driver behavior detection system in intelligent transportation systems. IEEE Veh. Technol. **62**(9), 4264–4275 (2013)
11. Johnson, R.R., Popovic, D.P., Olmstead, R.E., et al.: Drowsiness/alertness algorithm development and validation using synchronized EEG and cognitive performance to individualize a generalized model. Biol. Psychol. **87**(2), 241–250 (2011)
12. Touryan, J., Lance, B.J., Kerick, S.E., et al.: Common EEG features for behavioral estimation in disparate, real-world tasks. Biol. Psychol. **114**, 93–107 (2016)
13. Yang, G., Lin, Y., Bhattacharya, P.: A driver fatigue recognition model based on information fusion and dynamic Bayesian network. Inform. Sci. **180**(10), 1942–1954 (2010)

14. Mohamed, F., Nataraj, S.K., Ahmed, S.F., et al.: An approach in determining fatigueness and drowsiness detection using EEG. Res. Inventy Int. J. Eng. Sci. **8**(3), 20–28 (2018)
15. Mohamed, F., Ahmed, S.F., Ibrahim, Z., et al.: Comparison of features based on spectral estimation for the analysis of EEG signals in driver behavior. In: 2018 International Conference on Computational Approach in Smart Systems Design and Applications (ICASSDA). pp. 1–7, Conference Proceedings (2018)
16. Levenberg, K.: A method for the solution of certain non-linear problems in least squares. Q. Appl. Math. **2**(2), 164–168 (1944)
17. Marquardt, D.W.: An algorithm for least-squares estimation of nonlinear parameters. J. Soc. Ind. Appl. Math. **11**(2), 431–441 (1963)
18. Hagan, M.T., Menhaj, M.B.: Training feedforward networks with the Marquardt algorithm. IEEE T Neural Networ **5**(6), 989–993 (1994)
19. Boser, B.E., Guyon, I.M., Vapnik, V.N.: A training algorithm for optimal margin classifiers. In: Proceedings of the Fifth Annual Workshop on Computational Learning Theory, pp. 144–152, Conference Proceedings (1992)
20. Vapnik, V.N.: Statistical Learning Theory. Wiley, New York, US (1998)
21. Vapnik, V.N.: The Nature of Statistical Learning Theory. Springer Science & Business Media, Berlin (2000)
22. Brereton, R.G., Lloyd, G.R.: Support vector machines for classification and regression. Analyst **135**(2), 230–267 (2010)
23. Mitchell, T.M.: Artificial neural networks. Mach. Learn. **45**, 81–127 (1997)
24. Bahari, F., Janghorbani, A.: EEG-based emotion recognition using recurrence plot analysis and K nearest neighbor classifier. In: 20th Iranian Conference on Biomedical Engineering, ICBME 2013, pp. 228–233, Conference Proceedings (2013)
25. Yazdani, A., Ebrahimi, T., Hoffmann, U.: Classification of EEG signals using dempster Shafer theory and a K-nearest neighbor classifier. In: IEEE EMBS C Neur E, pp. 327–330 (2009)
26. Pan, J.-S., Qiao, Y.-L., Sun, S.-H.: A fast K nearest neighbors classification algorithm. IEICE T Fund Electr. **87**(4), 961–963 (2004)

Analytical Study of a Cylindrical Linear Electromagnetic Pulsing Motor for Electric Vehicles

N. M. Noor, Ishak Aris, S. Arof, A. K. Ismail, K. A. Shamsudin and M. Norhisam

Abstract The cylindrical linear electromagnetic pulsing motor (EMPM) is an alternative electric vehicle (EV) to be simulated in this study. The proposed design on the cylindrical linear EMPM will replace the piston engine in an internal combustion engine (ICE) which produces linear motion. It can eliminate problems related to internal combustion engines (ICE) such as engine weight and friction where fewer components have been used. In this paper, an analytical model was constructed and predicted the magnetic equivalent circuit (MEC) that can solve with the same technique as the electrical circuit. The initial magneto-statics analysis was conducted through the finite element magnetic software (FEMs) for magnetic filed problem so that the magnetic flux relationship could be predicted. Furthermore, the FE modelling and analysis is followed by a MATLAB/Simulink software calculation to predict the cylinder linear EMPM. Finally, the simulation results of the FE models regarding plunger force, thrust, plunger distance, speed, and power motor were presented and compared with the regulated counterparts obtained from the experimental setup.

Keywords Cylinder linear EMPM · ICE, MEC · FEMs · FE modelling

N. M. Noor (✉) · S. Arof · A. K. Ismail · K. A. Shamsudin
Electrical Electronic Automation Section, Universiti Kuala Lumpur, Malaysian Spanish
Institute, Kulim Hi-Tech Park, 09000 Kulim, Kedah, Malaysia
e-mail: noramlee@unikl.edu.my

S. Arof
e-mail: saharul@unikl.edu.my

A. K. Ismail
e-mail: ahmadkamal@unikl.edu.my

K. A. Shamsudin
e-mail: khairulakmal@unikl.edu.my

I. Aris · M. Norhisam
Department of Electrical and Electronics, Faculty of Engineering, University Putra Malaysia,
Jalan Upm, 43400 Serdang, Selangor, Malaysia
e-mail: ishak_ar@upm.edu.my

M. Norhisam
e-mail: norhisam_m@upm.edu.my

© Springer Nature Switzerland AG 2019
M. H. Abu Bakar et al. (eds.), *Progress in Engineering Technology*,
Advanced Structured Materials 119, https://doi.org/10.1007/978-3-030-28505-0_6

1 Introduction

In the last decade, the automobile industry is currently facing difficult issues such as global warming and a shortage of fossil fuel resources. Most vehicles on the road today use internal combustion engines (ICE) which convert thermal energy into mechanical energy. Statistics show that about 28% of the total air pollution is produced from combustion of petrol by ICE vehicles [1]. A significant of the total power loss in the current ICE is caused by the compression of the piston ring and cylinder wall, which is estimated at 35% of the overall mechanical friction machine [2]. They are many types of research and development of next-generation vehicles, which are extending from different angles [3]. The electric vehicle (EV) is the most important approach for environmental reasons. Therefore, to move an EV requires an electric motor which works on electromagnetic principles by converting the electrical energy into kinetic energy [4, 5]. There is an increased interest in vehicles using electric motors for driving power [6] such as induction motors (IMs) and permanent magnet synchronous motors (PMSMs). The evolution of electric motors has been moved one-step forward by introducing a linear motor with improved dynamic performance and reliability.

A linear motor for any electromagnetic tool that develops a mechanical thrust without the need for any gear or rotary device. Among the advantages of using a linear motor include the quieter operation, lower operating costs, faster processing, more accurate positioning, and longer life, less maintenance; fewer moving parts and multiple activities for features without gear [7, 8]. The design of the electromagnetic actuator was perform using 2D or 3D engineering finite element calclucation. This technique allows accurate device performance determination but requires large calculations, irrespective of time-reduction methods. The software requires an analytical description of the device operation. All this needs to be resolved by the model based on our model of electromagnetic reluctance and MMF.

The main focus of this research is to design, developed and analytically study of a new linear motor for EV application, as it is known as the cylindrical linear EMPM that creates a linear movement. Initially, the combination of a mechanical link between the cylindrical linear EMPM and crankshaft assembly was replaced by a piston engine so that the working principle was equivalent to the reciprocating motion in the ICE. An analytical model was constructed and predicted a MEC that can solve the task with the same technique as the electrical circuit. The initial magneto-statics analysis was conducted through the FEMM software so that the magnetic flux relationship could be predicted. Furthermore, the FE modelling and analysis is followed by a MATLAB/Simulink software calculation to predict the cylindrical linear EMPM. Finally, the simulation results of the FE models regarding plunger force, thrust, plunger distance, speed, and power motor are evaluated and compared to measured counterparts obtained from the experimental setup.

2   Cylinder Linear EMPM Structure

The structure of the cylindrical linear EMPM is like a linear motor or solenoid actuator. It contains several components such as york, coil winding, and plunger rod. A CAD software, CATIA V5, was used to design the overall components of the linear EMPM and the details of the design are given in Fig. 1. The mild steel AISI 1010 according to the American Iron and Steel Institute was used to fabricate linear EMPM components due to its better magnetic properties. Moreover, this mild steel was selected because of its good magnetic properties which contain 8–13% of carbon and considering its cost, permeability, and availability [9].

The cylindrical linear EMPM is working if a high DC current is applied to the coils winding where it creates external flux lines due to the current that flows perpendicular to the flux. The linear EMPM thickening depends on the presence of a magnetic field wherein this study the thickness of the radial gap is 0.2 mm. Therefore, to avoid fluid leakage from entering the coil bobbin, the yoke that embedded the coil bobbin is a seal to the inner housing wall. Thus, it will generate an attractive and repulsive force to produce the thrust or torque that moves the plunger rod to the right and left direction.

3   Magnetic Circuit Model Analysis

In order to facilitate the analysis and modeling the characteristics of a cylindrical linear EMPM, the equivalent magnetic circuit method was used to describe the relationship between critical design parameters and machine performance [10–13].

3.1   Predicted Magnetic Equivalent Circuit

The magnetic circuit model has been developed by plotting the estimated magnetic flux path in the air gap shown in Fig. 2a, i.e. the divided area for the equivalent

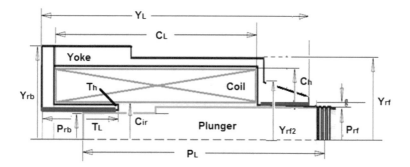

Fig. 1 Cross-section of the cylindrical linear EMPM components

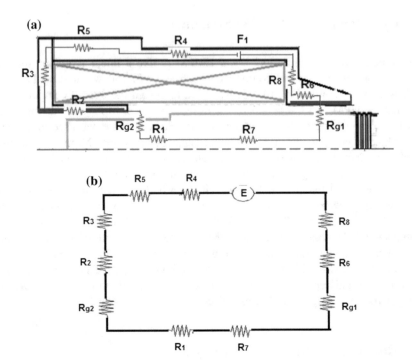

Fig. 2 **a** The divided area for equivalent magnetic circuit; **b** equivalent magnet resistance circuit

magnetic circuit situation and the situation taking the magnetic leakage conductance into account.

Figure 2b shows the equivalent magnetic resistance of the inside area of a cylinder linear EMPM model. The magnetic resistance from the proposed model can be describe as shown in Fig. 2a, b; each part of the resistance can be calculated as below:

$$R \int \frac{dx}{\mu A} \tag{1}$$

where R = reluctance.

According to the equivalent magnetic circuit shown in Fig. 3a, b, the magnetic flux, the magnetic conductance, and the leakage magnetic resistance can be derived as follows:

$$R_1 = \frac{P_L}{\mu \pi \times P_{rb}^2} \tag{2}$$

$$R_2 = \int\limits_{P_{rb}+g}^{P_{rf}} \frac{1}{2\pi \mu T_L} dr = \frac{\ln P_{rf} - \ln(P_{rb}+g)}{2\pi \mu T_L} \tag{3}$$

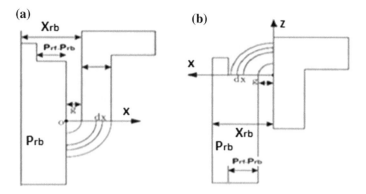

Fig. 3 Main magnetic flux paths. **a** Equivalent axial length front; **b** equivalent axial length back

$$R_3 = \frac{Y_L}{\mu\pi\left[Y_{rb}^2 - (P_{rb} + g)^2\right]} \tag{4}$$

$$R_4 = \frac{Y_L}{\mu\pi\left[Y_{rb}^2 - (P_{rf} + g + C_h + 0.1)^2\right]} \tag{5}$$

$$R_5 = \frac{Y_L}{\mu\pi\left[Y_{rb}^2 - (P_{rf} + g + C_h + 0.1)^2\right]} \tag{6}$$

$$R_6 = \int_{P_{rf+g}}^{P_{rf2}} \frac{1}{2\pi\mu(Y_L - C_L)}\,dr = \frac{\ln P_{rf2} - \ln(P_{rf} + g)}{2\pi\mu(Y_L - C_L)} \tag{7}$$

$$R_7 = \frac{P_L}{\mu\pi \times P_{rf}^2} \tag{8}$$

$$R_8 = \frac{Y_L}{\mu\pi\left[Y_{rb}^2 - (P_{rf} + g)^2\right]} \tag{9}$$

where μ_0 is the permeability of air, g is the air gap, P_L is the length from the plunger rod, P_{rb} is the radius back from plunger rod, P_{rf} is the radius front from the plunger rod, T_L is the teeth length from yoke, Y_L is the length yoke, Y_{rb} is the inner radius back from yoke, C_h is the width coil and C_L is the lenght coil.

Total resistance (R_t) from Fig. 2b can be represented as below:

$$R_t = R_1 + R_2 + R_3 + R_4 + R_5 + R_6 + R_7 + R_8 + R_{g1} + R_{g2} \tag{10}$$

The main magnetic flux paths are described in Fig. 3.

Refer to the main magnetic flux paths in Fig. 3, wherein Fig. 3a the equivalent axial length front as $X_{rb} - 2g + x$, and Fig. 3b indicates the equivalent axial length back as $X_{rb} - z$, The main magnetic flux can be calculated as follows.

where R_g is the reluctance gap

$$R_{g1} = \int_0^{d1 + d2 + 2g - move} \frac{\mu_0 \pi (X_{rb} - g)}{g} dz$$
$$= \frac{\mu_0 \pi (X_{rb} - g)}{g} (d1 + d2 + 2g - move) \tag{11}$$

$$R_{g2} = \int_0^{2g + move} \frac{\mu_0 \pi (X_{rb} - g)}{g} dz = \frac{\mu_0 \pi (X_{rb} - g)}{g} (2g + move) \tag{12}$$

3.2 Predicted Magnetic Analysis

The simulation of the magnetic flux density using FEMs is performed to estimate the presence of the magnetic field in the active area of the proposed cylindrical linear EMPM. Generally, the appearance of the magnetic field is difficult to measure experimentally and needs to be simulated used FEMs to predict the magnetic flux distribution, flux flow and mesh. In this model, structural and thermal responses are neglect. The magnetic circuit analysis is a simple method to estimate the cylindrical linear EMPM that can be provided by the fluid. The assigned materials for the yoke and plunger rod is low carbon steel (ANSI 1010) because of it's good magnetic penetration. In order to prevent the magnetic copper wire from contacting the steel, the non-magnetic stainless steel type 314 is used for the coil bobbin. In this study, several variables need to be assigned to conduct the simulation. The coil is made using copper wire (type 12 AWG) and the diameter of the coil is 4 mm with 858 turns; thus the result is 0.33 Ω coil resistance from the magnetic analysis. In this analysis, the supplied current is 200 A and 13.2 kW of cylindrical linear EMPM power consumptions. Next, the cylindrical linear EMPM is modeled in FEMs using 3D-axis-symmetry by selecting triangular elements. The meshing process was applied to the appropriate parts and completed the simulation as shown in Fig. 4. In the meshing process, all the parts were divided into small elements and nodes, and each element individually represented the material properties. Thus, the size of the mesh was a significant factor in this simulation to ensure the accuracy of the simulation results. The magnetic analysis was implemented to obtain the simulation result.

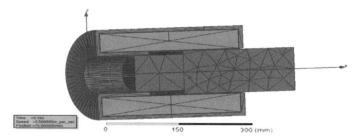

Fig. 4 A 3D Maxwell model of the meshing modeling environment

For this study, the magnetic flux density, magnetic flux flow path, and the force of direction were considered. This arrangement makes the flux pass radially through the air gap to the yoke teeth. The results are shown in a contrasting colour and spectrum. Each colour represents a spectrum of different intensities. For example, the red colour of the magnetic flux density has the highest value. The area of the model focuses on the red colour which means that specific fields would be saturated first, and this condition should be avoid. Thus, the blue colour represents the lower flux density. Usually, the blue region occurs in the air. Figure 5a shows the magnetic flux distribution and Fig. 5b shows the magnetic flux flow in the linear EMPM model. The highest magnetic flux density occurred at the radial gap and is 1.718 T when 200 A current are applied.

4 Experimental Setup

The experimental setup shown in Fig. 6 consists of the linear EMPM, speed sensor, flywheel, load cell, sensor encoder system and controller system. The connecting rod of the linear EMPM, crankshaft, disc sensor encoder, and flywheel are fixed to the movable jig table where they are kept in the initial position and directly connected to the load cell. The battery is the main source of electricity in the measurement setup where it supplies high current or voltage to control the linear EMPM. The current supply will be through the winding coil terminals where the action of the magnetic field current is in the radial direction, and it will produce a higher torque or torque so that it can move the plunger rod from the up and down movements. This movement will cause an exchange of mechanical energy in the crankshaft assembly where the linear EMPM will produce the rotational motion. Besides, the experiment will collect and measure the data such as output thrust or torque, plunger force, speed, plunger distance, and power of the motor.

(a)

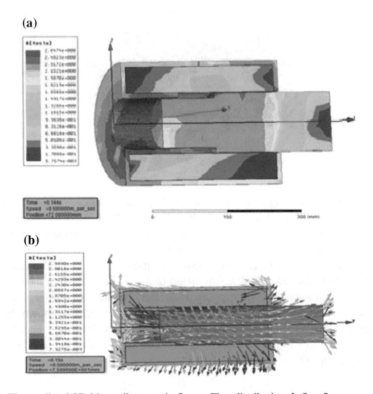

(b)

Fig. 5 The predicted 3D Maxwell magnetic flux. **a** Flux distribution; **b** flux flow

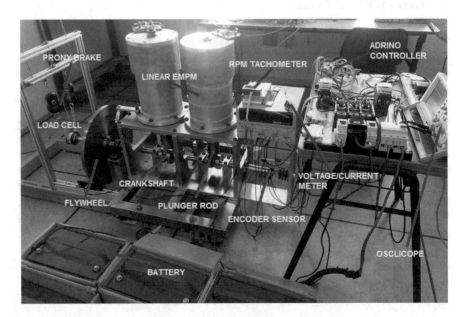

Fig. 6 Experimental setup for linear EMPM assembly evaluation

5   Results and Discussion

The FE modelling of the cylindrical linear EMPM model was simulated in the MATLAB/Simulink tool as shown in Fig. 7. The resistance (R) is 0.91 Ω, Inductance (L) is $2.543e^{-6}$ H.

5.1   Plunger Force

The graph in Fig. 8a shows the relationship between the plunger force and the switching time at a constant pulsing current is 200 A supply. The maximum plunger force of a cylindrical linear EMPM model is approx 37.2 kN. The graph in Fig. 8b shows the relationship between the plunger force and pulsing current. The plunger force is proportional to the pulsing current, therefore when the current is rising, the plunger power also increases rapidly.

5.2   Thrust

The graph in Fig. 9a shows the relationship between the thrust and the switching time for a constant pulsing current is 200 A supply. The thrust of a cylindrical linear EMPM model is approx. 60 Nm with the +ve pulse is activated. The graph in

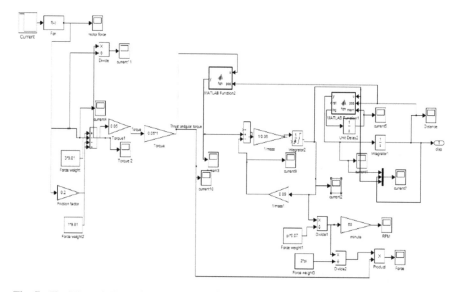

Fig. 7 The FE modelling of the cylindrical linear EMPM model

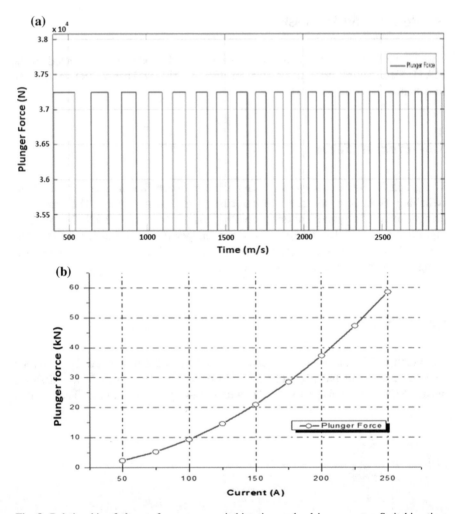

Fig. 8 Relationship of plunger force versus switching time and pulsing current. **a** Switching time; **b** pulsing current

Fig. 9b shows the relationship between the thrust and pulsing current. The thrust is proportional to the pulsing current, therefore when the current is rising, and the plunger power also increases rapidly.

5.3 Plunger Distance

The graph in Fig. 10a shows the relationship between the plunger distance and the switching time. The plunger distance is a constant value when the pulsing current is

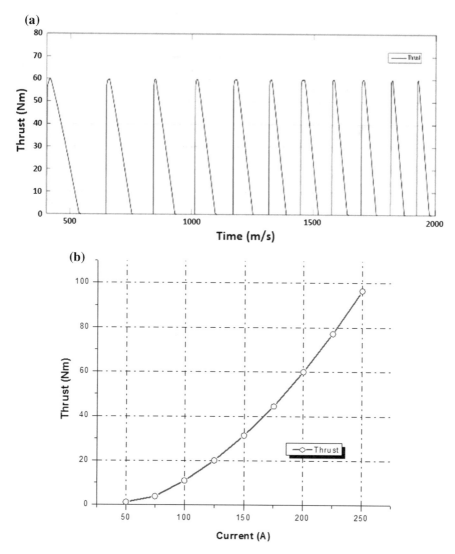

Fig. 9 Relationship of thrust versus switching time and pulsing current. **a** Switching time; **b** pulsing current

rising, and the maximum distance is 75 mm. The plunger distance of a cylindrical linear EMPM model is oscillation and the upper movement state of the crankshaft activated due to the plunger force and thrust designed to supplied is approx. 60 Nm. The graph in Fig. 10b shows the relationship between the plunger distance and pulsing current. Therefore, when the current increases, the plunger distance is a fixed value of 75 mm.

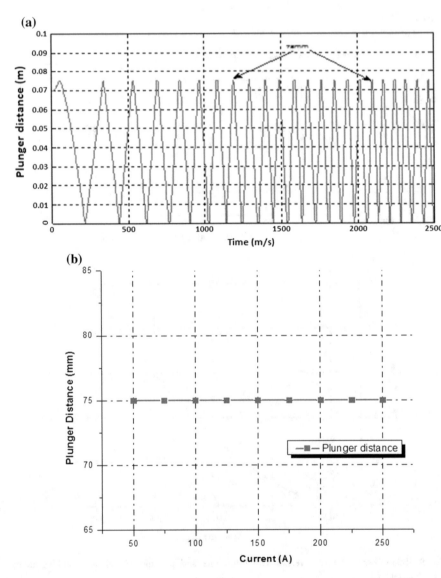

Fig. 10 Relationship of plunger distance and pulsing current. **a** Switching time; **b** pulsing current

5.4 Speed

The graph in Fig. 11a shows the relationship of speed and the switching time at a constant pulsing current supply of 200 A supply. The maximum speed of a cylindrical linear EMPM model is approx 4350 rpm which is the completed cycle

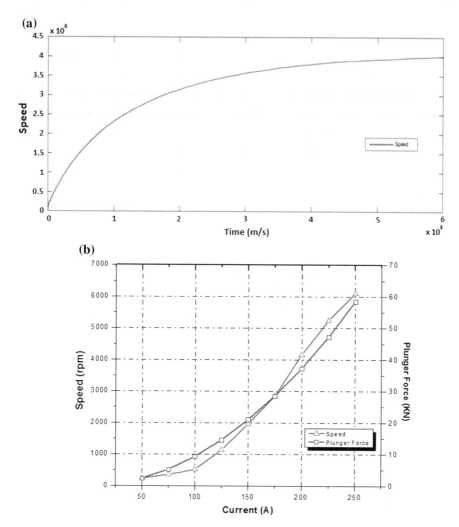

Fig. 11 Relationship of speed versus switching time and pulsing current. **a** Switching time; **b** pulsing current

of the plunger moving constantly. The graph in Fig. 11b shows the relationship between the speed to the plunger force and the pulsing current. The speed with plunger force is proportional to pulsing current, therefore when the current is rising, the plunger power also increases rapidly.

5.5 *Power Motor*

The graph in Fig. 12a shows the relationship between the power output motor and the switching time at constant pulsing current supply of 200 A. The power out of a cylindrical linear EMPM model is approx 6.4 kW. The graph in Fig. 12b shows the relationship power of the motor and the pulsing current. The power of the motor is proportional to the pulsing current, therefore when the current is rising, the plunger power also increases rapidly.

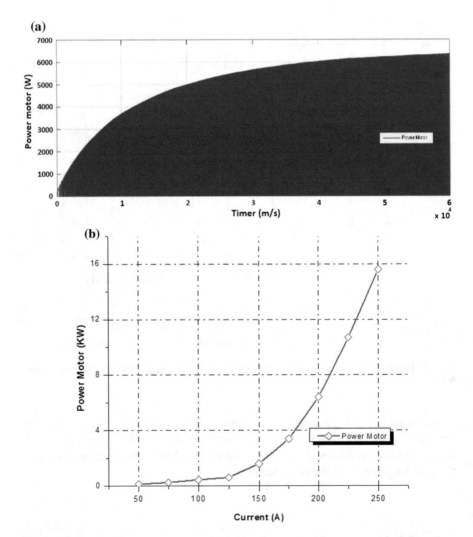

Fig. 12 Relationship power motor versus switching time and pulsing current. **a** Switching time; **b** pulsing current

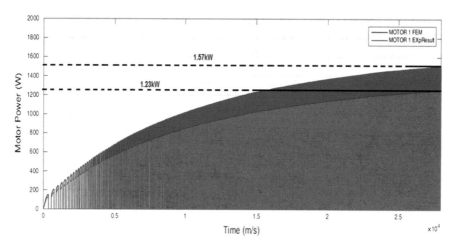

Fig. 13 Relationship of power motor versus switching time and pulsing current

5.6 Comparison Between Experimental Result and FEM Data

The graph in Fig. 13 shows the relationship between the power of the motor and the switching time with constant pulsing current is 150 A supply. The graph shows that the power of the motor based on the FEM analysis is approx. value is ~ 1.5 kW while the experimental result is approx 1.23 kW for cylindrical linear EMPM. Therefore, the comparison between the FEM/experimental power motor is 15% different. The pattern of the motor power graph in Fig. 13 is quite similar and increases rapidly.

6 Conclusions

In this paper, the main outcome of this work can be stated as follows listed and discussed such as; a cylindrical linear electromagnetic pulsing motor (EMPM) can be used in EV to replace the ICE system. In this paper, an analytical model was constructed and predicted equivalent magnetic circuit (MEC) that can solve the same technique as the electrical circuit. The initial magnetostatics analysis was conducted through the finite element method magnetics (FEMM) software so that the magnetic flux relationship could be predicted using Maxwell software. Furthermore, the FE modeling and analysis is followed by a MATLAB/Simulink software calculation to predict the cylindrical linear EMPM. The simulation results of the FE models are measure regarding plunger force, thrust, plunger distance, speed, and power motor are presented; and were compared with the measured counterparts obtained from the experimental setup and the test analysis are presented. Based on the analysis the system can be fabricated and tested in the future by considering the parameters that have been studied in this paper.

Acknowledgements The authors gratefully acknowledge financial support for this work by the University Putra Malaysia and Universiti Kuala Lumpur Malaysian Spanish Institute (Research Lab).

References

1. Andersson, B.S.: Company perspectives in vehicle tribology. In: Volvo 17th Leeds Lyon Symposium on Tribology Vehicle Tribology, Tribology Ser., Elsevier, 18, pp. 503–506 (1991)
2. Bolander, N.W., Steenwyk, B.D., Sadeghi, F., Gerber, G.R.: Lubrication regime transitions at the piston ring-cylinder liner interface. Proc. IMechE Part J: Eng. Tribol. **219**, 19–31 (2005)
3. Miler, J.M.: Hybrid electric vehicle propulsion system architectures of the e-CVT type. IEEE Trans. PLS **21**(3), 756–767 (2006)
4. Chan, C.: An overview of electric vehicle technology. Proc. IEEE **81**(9), 1202–1213 [Online] (1993, Sept). Available: http://ieeexplore.ieee.org/iel1/5/6088/00237530.pdf
5. Shimizu, H., Harada, J., Bland, C., Kawakami, K., Chan, L.:. Advanced concepts in electric vehicle design. IEEE Trans. Ind. Electron. **44**(1):14–18 [Online] (1997, Feb). Available: http://ieeexplore.ieee.org/iel1/41/12169/00557494.pdf
6. Jeong, Y., Sul, S., Schulz, S.E., Patel, N.R.: Fault detection and fault-tolerant control of interior permanent-magnet motor drive system form electric vehicles. IEEE Trans. IA **41**(1), 46–51 (2005)
7. Panahi, I., Arefeen, M., Yu, Z.: DSPs excel in motor-control applications. In: Electronic Design News (EDM) Magazine, R. Caves, Multinational Enterprise and Economic Analysis. Cambridge: Cambridge University Press (1997, 2008)
8. Lee, J.Y., Hong, J.P., Kang, D.H.: Analysis of permanent magnet type transverse flux linear motor by coupling 2D finite element method on 3D equivalent magnetic circuit network method. IEEE Trans. Magn. **5**, 2092–2098 (2004)
9. Sgobba, S.: Physics and measurements of magnetic materials, **4**, 25 (2011)
10. Pompermaier, C., Haddad, K.F.J., Zambonetti, M.V.A., da Luz, F., Boldea, I.: Small Linear PM oscillatory motor: magnetic circuit modeling corrected by axisymmetric 2-D FEM and experimental characterization. IEEE Trans. Industr. Electron. **59**(3), 1389–1396 (2012)
11. Sheikh-Ghalavand, B., Vaez-Zadeh, S., Hassanpour Isfahani, A.: An improved magnetic equivalent circuit model for iron-core linear permanent-magnet synchronous motors. IEEE Trans. Magn. **46**(1), 112–120 (2010)
12. Chillet, C., Voyant, J.Y.: Design-oriented analytical study of a linear electromagnetic actuator by means of a reluctance network. IEEE Trans. Magn. **37**(4), 3004–3011 (2011)
13. Zhu, Z.Q., Chen, X., Howe, D., Iwasaki, S.L.: Electromagnetic modeling of a novel linear oscillating actuator. IEEE Trans. Magn. **44**(11), 3855–3858 (2008); Lonngren, K.E., Bai, E.: On the global warming problem due to carbon dioxide. Energ. Policy **36**(4), 1567–1568 (2008). https://doi.org/10.1016/j.enpol.2007.12.019

Investigation on Effective Pre-determined Time Study Analysis in Determining the Production Capacity

Mohd Norzaimi Che Ani and Ishak Abdul Azid

Abstract In this paper, the application of pre-determined time study analysis in determination of production capacity had been investigated to understand the accuracy level of production capacity. In determination of production capacity, normally motion time study analysis is conducted for actual production processes to define the bottleneck process and production capacity. However, for new product introduction (NPI), the calculation of the labor cost and the unfamiliar production process is normally based on assumptions or benchmarking from similar processes. As introduced by Maynard Operation Sequence Technique (MOST®), it helps the industries in determining the production capacity. The dilemma of industries is the level of accuracy for a pre-determined time study at the beginning stage of production. This research has been conducted by examining the accuracy of pre-determined time study using the MOST® technique in selected case study industries and the results of this study show that the level of accuracy achieved is at 83.84%.

Keywords Pre-determined time study · Maynard operation sequence technique (MOST®) · Bottleneck process · Motion time study · Production capacity

M. N. Che Ani (✉)
Manufacturing Section, Universiti Kuala Lumpur Malaysian Spanish Institute,
Kulim Hi-Tech Park, 09000 Kulim, Kedah, Malaysia
e-mail: mnorzaimi@unikl.edu.my

I. Abdul Azid
Mechanical Section, Universiti Kuala Lumpur Malaysian Spanish Institute,
Kulim Hi-Tech Park, 09000 Kulim, Kedah, Malaysia
e-mail: ishak.abdulazid@unikl.edu.my

© Springer Nature Switzerland AG 2019
M. H. Abu Bakar et al. (eds.), *Progress in Engineering Technology*,
Advanced Structured Materials 119, https://doi.org/10.1007/978-3-030-28505-0_7

1 Introduction

The motion time study of the production floor was one of the effective tools to determine the bottleneck process and to identify the production capacity. In the production system, the entire process cycle times for each motion were reviewed and the highest process cycle time was determined as the bottleneck process because that process would determine the production output. A motion time study as an important tool in investigates human work in all its contents, which leads to systematic investigation of all the factors that affect the efficiency of the situation being reviewed, in order to seek improvement [1]. Normally, the tool of studying the process cycle time in a motion time study analysis was by a using stop watch which could be used to identify the process cycle time of workers in performing the tasks. This required several steps such as identifying step-by-step of motions performing the tasks, capturing the motions time with several trials, recording into the check sheet and calculating the average from the captured time.

The challenging of determination of the production capacity using the motion time study is for the new product introduction (NPI). Since the physical setting of the production layout was still not available in the phase of NPI, the pre-determined time study known as Basic Maynard Operation Sequence Technique (Basic MOST®) was applied [2]. The current scenario shows most of the industrial practitioner employed the MOST® technique for pre-determined time study and it is widely applied because it is more sophisticated than the Methods of Time Measurement (MTM) technique, and is also recognized as a global standard [3]. User-friendly and easy to learn, Basic MOST® has been accepted by countless industries as one of the most efficient work measurement techniques available [4].

The level of accuracy for pre-determined time studies at the beginning stage of the production should be tally with the actual production system. So, the dilemmas of the industries in determining of the production cycle time using Basic MOST® is the accuracy of the analysis to transform into the actual production system. Puvanasvaran [3] highlighted that the problem in computing the effective production system is the inaccuracy of the data used and a lack of a medium to evaluate the improvement ideas before it is implemented. Therefore, it is the duty of the industrial practitioner is to ensure the accuracy and the reliability in the determination of the production cycle time. Thus, the production capacity should be determined accurately to meet the customer order.

Thus, this research has been conducted by examining the accuracy of pre-determined time study using of the Basic MOST® technique. The first objective is to determine the cycle time of production processes using Basic MOST®. The second objective is to study the production cycle time using time study analysis (stopwatch) and the last objective is to examine the effectiveness of the Basic MOST® versus time study analysis. The research presented in this article had been divided into five main sections. The first section overviews the concept of time study analysis using Basic MOST®, then followed by second section which is the overview on how this research was conducted in the methodology section.

Completion of the third section, then data collection, analysis and discussions will be elaborated in the fourth section. The overall achievement of this research article will be concluded in Sect. 5.

2 The Background of Pre-determined Time Study

Saito [5] states various methods to identify the production capacity using the time study technique in his article which needs to be applied on the process in any organization. A time study analysis is considered as an internal audit in a production system to review the performance of the workers at machines in terms of cycle time to complete the tasks and a comparison with the planned capacity will be conducted. Normally, during planning the production capacity in the beginning stage some estimation has been used such as the estimation of cycle time. Based on this scenario, few literature has been identified for optimizing the productivity using the time study analysis [6, 7], continuous improvement [8, 9] and optimization of assembly line [10, 11] in manufacturing industry was analyzed and the results show that the accuracy of the process estimation is important in determination an effective production system.

From all of these articles, fewer empirical references in term of accuracy of pre-determined time study were found [12]. Various work measurement techniques in the time study analysis were discussed and for improvement of process and labour productivity which is a work measurement technique that basically concentrates on the mass production system [13]. Basic MOST® is not that much popular for doing research as compared to other industrial techniques and only few references have discussed [4, 14, 15] that on how Basic MOST® can be used for improving productivity and establishing time standards for a particular process mainly in manufacturing organization. Rane [16] discussed about the complexity of the assembly process for vehicle industries and this study also stated that for improving performance in industries is crucial, but industries still follow techniques of actual time study using stop watches. Thus, the implication of application the Basic MOST® in determination the production capacity needs to be considered as the main element in order to ensure the effectiveness of the production system and the accuracy of the Basic MOST® must be further investigated.

3 The Research Framework

This section presents and discusses the application framework of investigation on effective pre-determined time study analysis in determining the production capacity through a real world case study implementation. In order to determine the accuracy of the pre-determined analysis, it is imperative to comprehensively determine the actual steps of the study itself. From the developed framework, in this study, three

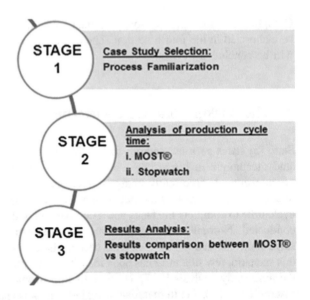

Fig. 1 The research framework in determination the accuracy of pre-determined time

main phases were identified which are identification of the selected case study to conduct pre-determined time study in phase 1, then followed by the analysis of the time study using stop watch and Basic MOST® technique in the second phase, and then results comparison between MOST® versus stopwatch will be analyzed to determine the accuracy of the pre-determined time study in the last phase. The application of the research framework in determination the accuracy of the pre-determined time study is summarized in Fig. 1.

4 Data Collection, Analysis and Discussions

The production processes have been observed by conducting the pre-determined time study and the actual process cycle time is determined by a using stop watch at the two different case study industry to determine the accuracy of pre-determined time study. The production processes were observed and analyzed based on the developed framework as discussed in the previous section.

4.1 Case Study Selection: Process Familiarization

The first case study was selected from a vendor of automotive industry which involved manual assembly during performing the production process. The selected company supplies the engine electronic control boards, instrument clusters and timing devices to the reputable automotive manufacturers. The production shop floor consists of three different sections: part preparation, surface mounting process and back end. The process is sequential, in batch, multi-path and asynchronous. At the back end, there are several assembly cells, each accommodates different product families. The focus of the case study is Cell-A, one of the cells producing mixed models of the instrument display. The processes involved depanelling, dip soldering, touch-up, in-circuit test, and functional visual test. Its layout is U-shaped with exit and entrance points product-dependent. A dedicated supermarket rack is available to store incoming materials pushed from the surface mounting process in batches. Consequently, the loading at Cell-A has to depend on the supplies from the surface mounting process. The outputs from Cell-A are sent to quality assurance.

The second case study produces the refurbishment of petroleum gas and refurbishment services for a liquefied petroleum gas (LPG) cylinder. This company provides refurbishment services for LPG cylinder in accordance to the specification and requirement of the customer. Generally, in the production layout the machines and equipment are arranged in a single line based on a product-based production system, which depends on the sequence of process activity and are connected and the material in the production floor typically flows from one workstation directly to the next workstation. The entire processes in the production system are human depending and a lot of manual work such as transferring the production between workstations, lifting, labelling, checking and crating is involved. Continuously monitoring and improvement of the quality system was applied in the working culture of the organization. With rapidly increasing demand in production, the selected case study companies need to increase their potential in production and effectiveness to compete against their competitors. At the same time, the production process needs to be equipped with the ability to have lower cost with higher effectiveness. The production floor is divided into two main sections known as front-of-line (FOL) and end-of-line (EOL), where the FOL is performing the cleaning, weighing, welding and re-painting tasks. While, in EOL, the semi-finished product from FOL will be continued with finishing processes such as drying, labelling and weighing conformation prior transfer into transportation for delivery process.

4.2 Analysis of the Production Cycle Time: Basic MOST® Versus Stop Watch

From the selected case study industries, then the production cycle times were analyzed using Basic MOST® and the actual time study using a stop watch. Both of the techniques have been used to determine the accuracy of the pre-determined time study.

Table 1 Production cycle time analysis using Basic MOST® for case study 1

No.	Method description	Activities sequences	TMU
1	Remove plastic from the dial	A1B0G1M3X0I0A0	50
2	Place dial into the housing	A0B0G0A1B0P6A0	70
3	Remove plastic from backcase	A1B0G1A1B0P0U3A1B0P1A0	80
4	Place backcase into the fixture	A0B0G0A1B0P3A0	40
5	Place PCB assy into backcover	A1B0G1A1B0P3A0	60
6	A-side finished	A1B0G1A1B0P1A0	40
Total TMU			340
Total time in seconds			12.27

4.2.1 Case Study 1: The Vendor of Automotive Industry

Based on the selected processes in the case study industry, the Basic MOST® has been applied to determine the production cycle time. The analysis using Basic MOST® was thoroughly analyzed based on detail motion elements of selected processes and the results of cycle time was obtained as 12.24 s as shows in Table 1.

Successful analyzed time study by using Basic MOST®, with the same selected process of production, the actual time study by using stop watch was performed. In this stage, the identification of the workers in performing the tasks was selected. The workers were selected from those who were in-charge of that process in the same production line. Thus, the analysis of the motion study was conducted based on five repetitive motions. Then, the average time should be calculated from the data collection to ensure the accuracy of the data collected of the five repetitive motions. The chosen measurement technique was manual observation with time taking by using a stop watch because it reflected the actual situation. Based on the repetitive process, the summary of time recorded based on average is tabulated in Table 2. The data from Table 2 shows that the amount of time required was 14.77 s.

Table 2 Production cycle time analysis using actual time study (stop watch) for case study 1

No.	Method description	Seconds					
		1st	2nd	3rd	4th	5th	Ave.
1	Remove plastic from the dial	1.21	1.88	1.00	1.85	1.50	1.49
2	Place dial into the housing	3.00	3.10	3.88	3.51	3.56	3.41
3	Remove plastic from backcase	2.50	2.66	2.85	2.61	3.86	2.90
4	Place backcase into the fixture	2.00	2.41	2.67	2.81	3.00	2.58
5	Place PCB assy into backcover	2.50	2.00	3.00	2.50	3.00	2.60
6	A-side finished	1.50	2.00	2.00	2.50	1.00	1.80
Total time in seconds							14.77

Table 3 Production cycle time analysis using Basic MOST® for case study 2

No.	Method description	Activities sequences	TMU
1	Collect LPG	A3B1G3A6B1P1A0	150
2	Chipping process	A3B3G6A3B6P1U3A6B3P1A0	360
3	Move LPG	A1B1G3A6B1P1A0	130
4	Checking series number	A3B1G3A6B1P1A0	150
5	Move LPG	A1B1G3A3B1P1A0	100
Total TMU			890
Total time in seconds			37.65

Table 4 Production cycle time analysis using actual time study (stop watch) for case study 2

No.	Method description	Seconds					
		1st	2nd	3rd	4th	5th	Ave.
1	Collect LPG	7.21	7.88	8.00	6.85	8.50	7.69
2	Chipping process	18.00	16.20	14.88	17.51	18.56	17.03
3	Move LPG	6.50	6.66	7.85	5.61	6.86	6.70
4	Checking series number	5.00	5.41	6.67	7.81	8.00	6.58
5	Move LPG	6.50	6.00	7.00	6.50	6.00	6.40
Total time in seconds							44.39

4.2.2 Case Study 2: The Refurbishment of Petroleum Gas and Refurbishment Services

Similar with the data collection in case study 1, in this case study 2 the same method has been applied which are determining the time study using Basic MOST® and actual time study by using a stop watch. The results of both analyses as shown in Table 3 for the analysis using Basic MOST® and Table 4 for the analysis using a stop watch. From the analysis using Basic MOST®, the obtained result was obtained 37.65 s, while by using a stop watch, the results shows 44.39 s.

4.3 Results Analysis: Comparison Between Basic MOST® Versus Stop Watch

From the results obtained from two different industries as discuses in previous section, then the overall results was analyzed in order to determine the level of accuracy of the pre-determined time study using Basic MOST®. The final results of both case studies as tabulated in Table 5 and shows the accuracy of case study 1 achieved 82.87% accuracy, while case study 2 achieved 84.82%. Averagely, the

Table 5 The results comparison between Basic MOST® versus stop watch

Case study	Method	Results	Accuracy (%)	Average accuracy (%)
1	Basic MOST®	12.24	82.87	83.84
	Stop watch	14.77		
2	Basic MOST®	37.65	84.82	
	Stop watch	44.39		

accuracy of determination of production cycle time using pre-determined time study is 83.84% based on implementation in two selected different case studies.

5 Conclusions

In this paper, an effective pre-determined time study analysis in determining the production capacity was investigated in selected case study industries to determine the accuracy of the Basic MOST® technique. In overall the results of this research have met the objectives as defined in prior conducting this research which are the investigation in selected case study industries was conducted to determine the effectiveness of pre-determined time study versus actual time study and 83.84% of accuracy was obtained. The future research of extension this research will focus on the customization of determination of time study analysis using pre-determined time study for individual selected industries.

Acknowledgements The authors also acknowledge the Universiti Kuala Lumpur, Malaysian Spanish Institute (UniKL MSI) for funding the study that resulted in publishing this article. Also highly appreciation extended for selected case study industry and anonymous reviewers for the comments and advises given which lead to the significantly improved the quality of this research article.

References

1. Adebayo, A.: an investigation into the use of work study techniques in Nigerian manufacturing organizations. Res. J. Appl. Sci. **2**(6), 752–758 (2007)
2. Yadav, T.K.: Measurement time method for engine assembly line with help of Maynard Operating Sequencing Technique (MOST). Int. J. Innov. Eng. Technol. (IJIET) **2**, 131–136 (2013)
3. Puvanasvaran, A., Mei, C., Alagendran, V.: Overall equipment efficiency improvement using time study in an aerospace industry. Proc. Eng. **68**, 271–277 (2013)
4. Gupta, M.P.K., Chandrawat, M.S.S.: To improve work force productivity in a medium size manufacturing enterprise by MOST Technique. IOSR J. Eng. **2**, 8–15 (2012)
5. Saito S (2001), Case study: reducing labor cost using industrial engineering techniques. In: Maynard's Industrial Engineering Handbook, Tokyo, Japan

6. Gorantiwar, V.S., Shrivastava, R.: Identification of critical success factors for quality-productivity management approach in different industries. Int. J. Product. Qual. Manage. **14**, 66–106 (2014)
7. Ani, M.N.C., Hamid, S.A.: Analysis and reduction of the waste in the work process using time study analysis: a case study. Appl. Mech. Mater. **660**, 971–975 (2014)
8. Singh, G., Singh, A.I.: An evaluation of just in time (JIT) implementation on manufacturing performance in Indian industry. J. Asia Business Stud. **8**, 278–294 (2014)
9. Ani, M.N.C., Ismail, A.B., Mustafa, S.A., et al.: Simulation analysis of rabbit chase models on a cellular manufacturing system. Appl. Mech. Mater. **315**, 78–82 (2013)
10. Ahmad, R., Kamaruddin, S.: An overview of time-based and condition-based maintenance in industrial application. Comput. Ind. Eng. **63**, 135–149 (2012)
11. Soroush, H., Sajjadi, S.M., Arabzad, S.M.: Efficiency analysis and optimisation of a multi-product assembly line using simulation. Int. J. Product. Qual. Manage. **13**, 89–104 (2014)
12. Jain, R., Gupta, S., Meena, M., Dangayach, G.: Optimisation of labour productivity using work measurement techniques. Int. J. Product. Qual. Manage. **19**, 485–510 (2016)
13. Neely, A., Gregory, M., Platts, K.: Performance measurement system design: a literature review and research agenda. Int. J. Oper. Prod. Manage. **25**, 1228–1263 (2005)
14. Rajvanshi, P.K., Belokar, D.R.: Improving the process capability of a boring operation by the application of statistical techniques. Int. J. Sci. Eng. Res. **3**, 1–6 (2012)
15. Tuan ST, Karim A, Kays H et al (2014) Improvement of workflow and productivity through application of Maynard operation sequence technique (MOST). In: Proceedings of the 2014 international conference on industrial engineering and operations management, vol 1, pp 7–9
16. Rane AB, Sudhakar D, Rane S (2015) Improving the performance of assembly line: review with case study. In: Nascent technologies in the engineering field (ICNTE) international conference, vol 1, pp 1–14

Vibration Measurement on the Electric Grass Trimmer Handle

**Muhammad-Najib Abdul-Hamid, Farahiyah Mahzan,
Shahril Nizam Mohamed Soid, Zainal Nazri Mohd Yusuf
and Nurashikin Sawal**

Abstract It is important to reduce the vibration level on the electric grass trimmer so that it is safe to be used by the user to avoid illness such as white fingers. The objective of this study is to measure the vibration level of an existing electric grass trimmer and to reduce the vibration level at the handle of the electric grass trimmer. The vibration level has been measured by two types of experiments which are spectral testing and impact test for modal analysis. A new handle has been designed by adding a spring stiffness in order to reduce the vibration level. Vibration level of the new handle is 0.22 g. After that, an active vibration control system (AVC) is developed using the LabVIEW system which applied a block diagram as its interface and solenoid as its actuator.

Keywords Hand arm vibration · Electric grass trimmer · Vibration level · Vibration measurement

M.-N. Abdul-Hamid (✉) · F. Mahzan · S. N. Mohamed Soid · Z. N. Mohd Yusuf · N. Sawal
Universiti Kuala Lumpur Malaysian Spanish Institute, Kulim Hi-Tech Park, 09000 Kulim,
Kedah, Malaysia
e-mail: mnajib@unikl.edu.my

F. Mahzan
e-mail: farahiyahmahzan@gmail.com

S. N. Mohamed Soid
e-mail: shahrilnizam@unikl.edu.my

Z. N. Mohd Yusuf
e-mail: zainalnazri@unikl.edu.my

N. Sawal
e-mail: nurashikins@unikl.edu.my

© Springer Nature Switzerland AG 2019
M. H. Abu Bakar et al. (eds.), *Progress in Engineering Technology*,
Advanced Structured Materials 119, https://doi.org/10.1007/978-3-030-28505-0_8

1   Introduction

Hand-arm vibration is a condition where vibration transmits from a tool or machine to a operator's hands and arms. Hand-arm vibration level is discovered by quantifying the vibration motion of the machine or tool held by the human. Meanwhile, the Hand-arm Vibration Syndrome (HAVS) is a disease that involves circulatory disturbances (example; vibration white finger), motor and sensory disturbances and musculoskeletal disturbances which may happen to the human who operates equipments that produce vibration [1]. A few studies related to hand-transmitted vibration show a very serious phenomenon. The human effects when exposed to high vibration level for a lengthy time are usually permanent and considered to be an occupational disease leading to invalidity [2]. Goglia reported that the whole-body vibration exposed to a framesaw user is at higher vibration level than the guidelines given in the new ISO 2631-1-1-1997.

In 2002, data from the European Commission showed that almost 17% of the European workers were exposed to vibration from machinery or handheld tools for at least half of their working time [3]. In order to protect the workers from HAVS, different countries have proposed and developed their own criteria and guidelines. For example, the European Union (EU) and Sweden National Institute for Working Life and OPERC database have formed the Human Vibration Directive and the Hand-Arm Vibration Database, respectively. The main objective is to measure the vibration level of handheld powered tools and discloses the HAVS effect to human body. Different frequency value of vibration will impede with different parts of human body where the whole body vibration happens at 1–30 Hz frequencies whereas segmental vibration which interferes with the hand-arm system is between 30 and 100 Hz [4]. When above 100 Hz, the human hand is affected by the vibration frequency [5]. Other findings propose a better range from 2 to 100 Hz for whole- body vibration and from 8 to 1500 Hz for segmental vibration and the current NIOSH guideline recommends the measuring of vibration up to 5000 Hz bandwidth [6].

An electric grass trimmer is used to trim the edge of lawn along the garden or other places that a mower cannot reach which also has general utility for trimming bushes, and hedges. An electric grass trimmer is simple and rugged in its construction, relatively light in weight and relatively low in cost. An electric grass trimmer also designed for quiet and zero carbon emissions and it's does not cause air pollution. Part of the electric grass trimmer consists of a powerful motor and also a string nylon blade. Both of these part will produce a high level of vibration during operation which will cause a forced vibration. The vibration will be transferred to the handle of the electric grass trimmer. Since the user is holding the handle during the operation of the electric grass trimmer, he or she can feel the vibration. However, the user may be unable to control and reduce the level of vibration. High level of vibration may lead to disease such as white fingers if the vibration is uncontrollable.

2 Methodology

The study started with the experimental phase. In this phase, a vibration measurement is conducted to find the vibration frequency and vibration level of the electric grass trimmer using spectral testing. The natural frequencies of handle and electric grass trimmer is also obtained using experimental modal analysis. After that, a new handle of electric grass trimmer was designed by adding a spring stiffness and material damping. After the design phase, this project is continued with the development of an active vibration control (AVC). During the development phase, the algorithm or the block diagram for the active vibration control is developed using the LabView software.

2.1 Vibration Measurement

Spectral testing was used to determine the vibration level of the electric grass trimmer. During the experiment, there were several points marked on the electric grass trimmer as the reference points to measure the vibration level. First of all, the electric grass trimmer was divided into five points as shown in Fig. 1. The distance of each point was measured using a measuring tape and has been marked. Vibration measurement was performed using LMS Data Acquisition System (DAQ) with accelerometer sensors. Then, accelerometer was mounted at the marked point on the electric grass trimmer. Before the measurement, the accelerometer was calibrated by using the accelerometer calibrator to identify the actual sensitivity.

Once the electric grass trimmer was switched ON, the accelerometer sensed the vibration level and recorded it by the DAQ. The measured data was then processed using Fast Fourrier Transform (FFT) to plot the amplitude of vibration versus frequency. From this graph, the maximum amplitude and vibration frequency can be identified. This step is repeated at each point with three different axis which are the x, y and z axis based on a basicentric coordinate system to determine which axis has the highest vibration level according to the hand-arm vibration direction.

2.2 Experimental Modal Analysis

Experimental modal analysis is a experimental technique to obtain dynamics characteristic of a structure which are the natural frequency, damping and mode shapes. Knowing system natural frequencies and mode shapes is key to avoid the

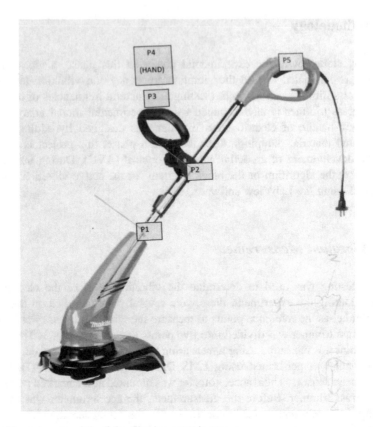

Fig. 1 Measurement points of the electric grass trimmer

resonance condition. Resonance is the excitation of a system at one of its natural causing vibration amplitudes to increase unbounded until failure comes about, whether due to fracture in the short term, or fatigue in the long term. Experimental modal analysis has been conducted at the rod and the handle of the electric grass trimmer as shown in Fig. 2. Each part needs to be set their geometry first.

For this experiment, an impact hammer for input 1 and an accelerometer for input 2 has been used. As the accelerometer is set as reference point at point 7 of the rod and point 4 of the handle. This experiment is conducted as the electric grass trimmer is switch OFF. The structure is being excite to have the vibrating responses. The impact hammer is used to excited the structure by knocking at the structure point to produce a short impact impulse duration. The impact hammer is coupled with a force sensor to measure the impact or excitation force that will be used to calculate the frequency response function (FRF).

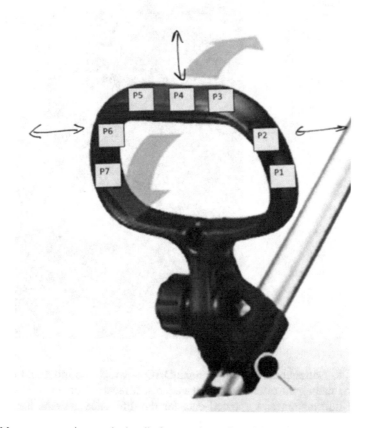

Fig. 2 Measurement points on the handle for experimental modal analysis

2.3 New Handle Design

The next phase is to design the new handle and to find the optimum value for mass, stiffness and damping. A new handle of the electric grass trimmer was designed with consideration to locate an actuator for the active vibration control. The new handle was designed with addition of a spring stiffness as shown in Fig. 3 so that the handle has stiffness and damping to absorb the vibration compared to the existing handle which was rigid.

2.4 Active Vibration Control

The AVC system was developed using the LabVIEW control design and simulation tools. The computer device that provided with the LabVIEW software was

Fig. 3 New handle of
electric grass trimmer

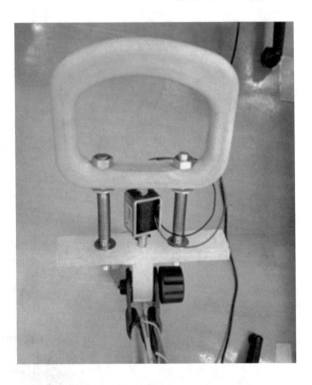

connected to National Instrument CompaqDAQ at which the input and the output unit are the main controller unit for all systems. It received data from the sensors, processed information and updated data for the difference systems and switched output devices. The output signal was used to activate the actuator as pushing the handle when the vibration level reached at a certain level.

This Active Vibration Control system consists of the input, process flow and output of the system. The block diagram is shown as in Fig. 4, the input consists of DAQ assistant which acts as the input to receive the signal from the sensor (accelerometer). The process is where the RMS value is simulated using the signal simulation to determine whether the command is true or false. When the RMS value is higher than the set trigger value, as for the output, it will give the voltage output to the solenoid after the process flow determine the command. Once the voltage output is given, the solenoid will push to the handle to reduce the vibration of the handle of the electric grass trimmer.

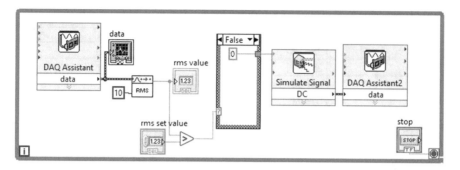

Fig. 4 Vibration control block diagram using LabVIEW

3 Results and Discussion

The result shows peaks of vibration level at several frequencies when the electric grass trimmer is operated as shown in Fig. 5. The peak amplitude occurs at each different point with different values. It is shown that the position of the measurement points affects the amplitude, g. This is because at P3 which is located at the handle has the greatest value of amplitude, g which is 5.90 g. This amplitude shows where and how the amplitude will act on the structure at any point of electric grass trimmer.

A high vibration amplitude occurs at P3 with 5.90 g at 251 Hz which is the operating speed for the blade. From the measurement, the dominant axis where vibration occurs is the X-axis based on the basicentric coordinate system as shown in Table 1.

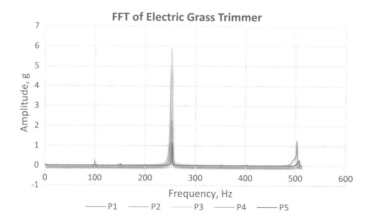

Fig. 5 Result of vibration measurement for electric grass trimmer with existing handle at various measurement points

Table 1 Vibration level at point measurements for X-axis

Point	Frequency (Hz)	Amplitude (g)
P1	251.17	4.82
P2	251.17	1.27
P3	251.88	5.9
P4	252.58	2.35
P5	254.00	1.10

Table 2 Natural frequencies for electric grass trimmer rod and handle

Part	Natural frequency (Hz)	Damping ratio (%)
Grass trimmer rod	1st mode: 2356	1.63
	2nd mode: 2559	2.97
Handle	1st mode: 1503	4.41

3.1   Natural Frequencies

Table 2 shows the result of natural frequencies and damping ratio for the grass trimmer rod and handle. The first natural frequency of the grass trimmer rod is 2356 Hz meanwhile for the handle, first natural frequency is 1503 Hz. From these results, it is shown that the vibration of the grass trimmer does not coincide with the natural frequency to produce the resonance condition since the operating frequency of the grass trimmer is 251 Hz.

3.2   Reduction of Vibration Level

As shown in Table 3, the vibration amplitude of the electric grass trimmer has the highest value at P3 which is at the handle of electric grass trimmer with an amplitude of 5.90 g. Therefore, a new handle was developed in order to reduce the vibration transmitted to the human body. Measurement was done using a new handle of electric grass trimmer and the vibration level was found to be 0.22 g at 273 Hz. After that, a solenoid actuator was applied to the new handle to test the functionality of the active vibration control system (AVC). An AVC system which

Table 3 Reduction of vibration level

		Frequency (Hz)	Amplitude (g)
Existing handle		251	5.90
New handle	Without solenoid	274	0.22
	With solenoid	277	0.20

only focused on the handle has been developing to reduce this amplitude using solenoid as the actuator. The measurement was done and it was found that the vibration level reduces to 0.20 g at the same frequency. It showed that the AVC system is functioning, but the efficiency of the actuator is low since the solenoid actuator used in this study has a low pushing force. It could be replaced with a high force actuator in order to reduce further the vibration level.

4 Conclusions

This study was conducted to reduce the vibration level of the handle of an electric grass trimmer to ensure the vibration level is at the safe amplitude for the user. The findings were based on the analysis of the data collected from the experiment that has been conducted by the researcher. It was found that there were different amplitude values at different points even though it is just slightly different. Besides, this was brought about by using a new design handle and a solenoid actuator in the Active Vibration Control system, the vibration level still can be reduced to the safe level. From the experiment, it can be concluded that low amplitude, g of vibration is important to ensure the user safety and to avoid a dangerous disease. From the results, it is shown that the vibration level is reduced from a maximum value of 5.9 g of amplitude to a minimum value of 0.20 of amplitude, g at which the total percentage of reduction is almost 97%.

Acknowledgements All the experiment and analysis conducted by Vibration and System Dynamics Laboratory, Universiti Kuala Lumpur Malaysian Spanish Institute, Kulim Kedah, Malaysia. Thanks to laboratory technician, Mr. Norsham Malek in assisting the vibration measurement.

References

1. Vergara, M., Sancho, J.L., Rodriguez, P., Perez-Gonzalez, A., et al.: Hand- transmitted vibration in power tools: Accomplishment of standard and users' perception. Int. J. Ind. Ergonom. **38**(9–10), 652–660 (2008)
2. Goglia, V., Grbac, I., et al.: Whole-body vibration transmitted to the framesaw operator. App Ergonom **36**, 43–48 (2005). https://doi.org/10.1016/j.apergo.2004.09.005
3. European Commission (2004) Work and health in the EU. A statistical portrait. Data 1994–2002, Luxembourg. EUROSTAT. ISBN 92-894-7006-2
4. Greenslade, E., Larsson, T.J., et al.: Reducing vibration exposure from hand-held grinding, sanding and polishing powertools by improvements in equipment and industrial processes. J. Saf. Sci. **25**(1–3), 143–152 (1997)
5. Snook, S.: The practical application of ergonomics principles. J. Occup. Health Saf. Australia New Zealand **9**(6), 555–563 (1993)
6. Taylor, J.S.: Vibration syndrome in industry: dermatological viewpoint. Am. J. Ind. Med. **8** (45), 415–432 (1985)

Low Harmonics Plug-in Home Charging Electric Vehicle Battery Charger Utilizing Multi-level Rectifier, Zero Crossing and Buck Chopper

Part 1: General Overview

Saharul Arof, N. H. N. Diyanah, Philip Mawby, H. Arof
and Nurazlin Mohd Yaakop

Abstract This paper focuses on developing a battery charger for electric car. A novel topology of a battery charger is proposed. Conventional rectifier has drawbacks in term of harmonic currents. This paper describes about a five level single-phase rectifier associated with zero crossings circuit and buck chopper with a control signal which draws a clean sinusoidal line current for the application of low harmonics plug in home charging Electric Vehicle battery charger. The MATLAB/Simulink results reveal the proposed Electric Vehicle battery charger performance compared to the conventional method.

Keywords Multi level inverter · Battery charger · Multi-level rectifier · Buck chopper · PWM · THD

S. Arof (✉) · N. H. N. Diyanah · N. Mohd Yaakop
Electrical Electronic Automation Section, Universiti Kuala Lumpur, Malaysian
Spanish Institute, Kulim Hi-Tech Park, 09000 Kulim, Kedah, Malaysia
e-mail: saharul@unikl.edu.my

N. H. N. Diyanah
e-mail: diyanahhisham94@gmail.com

N. Mohd Yaakop
e-mail: nurazlin@unikl.edu.my

S. Arof · P. Mawby
University of Warwick School of Engineering, Coventry CV47AL, UK
e-mail: p.a.mawby@warwick.ac.uk

H. Arof
Engineering Department, Universiti Malaya, Jalan Universiti, 50603 Kuala Lumpur,
Malaysia
e-mail: ahamzah@um.edu.my

© Springer Nature Switzerland AG 2019
M. H. Abu Bakar et al. (eds.), *Progress in Engineering Technology*,
Advanced Structured Materials 119, https://doi.org/10.1007/978-3-030-28505-0_9

1 Introduction

The emission of hydrocarbons does not only pollute the environment but also contributes to global warming which melts icebergs and increases the sea level. Using efficient Electric Vehicles (EV) and Hybrid Electric Vehicles (HEV) for transportation is one of the solutions to reducing global hydrocarbon emissions [1, 2]. The need of the clean vehicle cause continuous research study in AC and DC drive system which involves power drive train, electric motors and power converters or DC choppers [3–9]. To improve the expected electric vehicle performance, efforts in studying and improving the control technique, optimization using artificial intelligent, and testing with hardware and simulation software are required [10–16]. All electric vehicles including of DC drive EVs require battery charging and the charging operation can happen while the EV is moving or it is at standstill. The vehicle's inertia provides a source for charging while moving, while electricity provides power when the vehicle is not moving.

Connection to an electric power grid allows opportunities such as ancillary services, reactive power support, tracking of output from renewable energy sources, and load balance. A battery charger consists of a transformer, rectifier, and a buck or boost chopper [1, 4]. An excellent, efficient and reliable battery charger should have a high, being low cost, low volume and at low weight [7, 8]. Battery charger operation relies on its components, control technique and switching strategies [7, 8]. In general, the battery charger and its control algorithm are implemented using, microcontrollers, digital signal processors and integrated circuits [7, 8].

Two critical aspects of charging batteries are charging time and battery life. A decrease in the power factor due to an increase in the firing angle and relatively high on the harmonic currents are the inherent drawbacks of the conventional battery charger design. Due to that, multilevel converters are introduced in recent years to attain high power quality, low switching losses, high voltage capability, and better efficiency. The significance of this research work is that the adopted five level single phase shall contribute to an improvement of the power factor as well as a reduction in the total harmonic distortion (THD) [2, 8]. Thus, a combination of multi-level and buck chopper is suitable for electric car battery charger application.

2 Methodology

2.1 Proposed Battery Charger with Multi Level Rectifier

The basic type of battery charger that can operate to charge batteries is shown in Fig. 1. A conventional battery charger consists of two different converter circuits, which are the bridge rectifier and buck chopper.

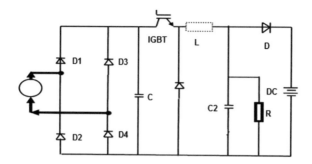

Fig. 1 Conventional battery charger with bridge rectifier and buck chopper

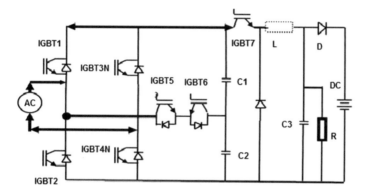

Fig. 2 Proposed battery charger

2.2 Proposed Multi Level Rectifier and Buck Converter for Battery Charging

The proposed battery charger utilizes a multi-level rectifier, its association with a buck converter and a zero crossing circuit to be used as a battery charger, see Fig. 2.

Combination of the proposed battery charger to DC drive four quadrants DC chopper will result in the configuration as shown in Fig. 3. This allows both charging at standstill at home and while the vehicle is moving whereby the four quadrants dc hopper provides power during generator mode.

2.2.1 Zero Crossing

Conventional zero crossing is a point where the sign of a mathematical function changes from positive to negative value or vice versa such as shown in Fig. 4.

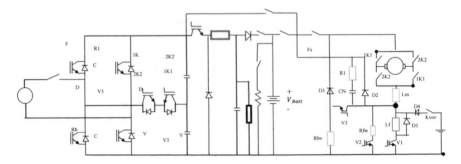

Fig. 3 Integrated FQDC chopper and battery charger

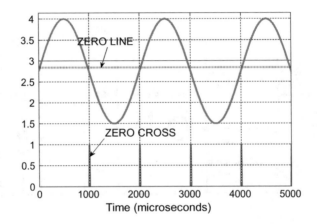

Fig. 4 Conventional zero crossing

Zero cross point from negative to positive as in Fig. 5 is the only used and important. Zero cross (ZC) detection has three functions, first it is to start the synchronization of the firing of carrier signal (triangular wave signal) and reference signals. Without zero cross the multilevel rectifier will not start at the exact be a point. Second, it is to point the start of the comparison process, there might be delay or offset between the both references (sine and carrier signal) at every zero cross point without it. Third, it is to refresh and reload the carrier signals value at every complete cycle. The action of the synchronization of the firing angle to incoming AC supply can result in a better performance (Efficiency, PF, THD) of MLR for the battery charger.

Zero crossing circuit requires a combination of a few circuits such as voltage sensor, offset circuit, and amplifier circuit. The voltage sensor is used to capture the AC voltage grid signal and to transform it to lower voltage which is safe and usable for control and signal conditioning purposes. An offset circuit is to offset the transformed AC voltage to complete DC voltage so that a digital controller such as a PIC microcontroller can be used to process the signals. Without offset circuit, the

Fig. 5 Right/required zero cross signal

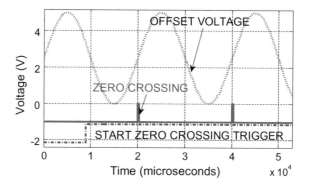

Fig. 6 Input and output signals of offset circuit

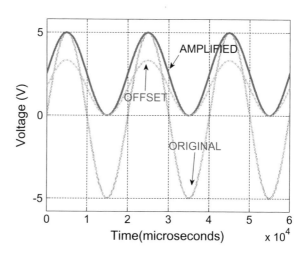

microcontroller is unable to read the negative part of the AC signal. Finally the amplifier is used to amplify the voltage to the desired value for the microcontroller. An offset circuit is shown in Fig. 6.

The algorithm begins when the start operate signal is received. As the required zero crossing for this circuit is from negative to positive, the Controller then will find the lowest point of input signal. Once this signal is obtained, the controller will produce the ready signal. As the sine input from the lower bottom is moving towards upwards, the controller will wait for the voltage to be in the range of half of the maximum peak of the voltage (Fig. 7).

The proposed battery charger consists of two different converter circuits, which are the multi-level rectifier and the buck chopper. The multi-level rectifier consists of a bridge rectifier and a multi-level inverter as shown in Fig. 8. Timer Interrupt (TU) and Look Up Table (LUT) techniques are used for this multi-level Rectifier IGBTs firing. Details about the firing technique are not discussed in this paper, but will be covered in another paper.

Fig. 7 ZC controller input
and output

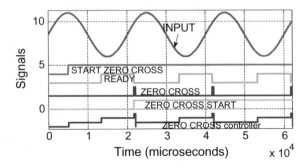

Fig. 8 Multilevel rectifier

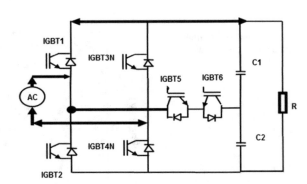

Fig. 9 Conventional bridge
rectifier circuit

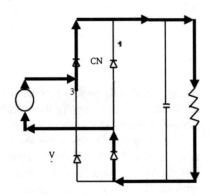

2.2.2 The Bridge Rectifier Circuit

The conventional bridge rectifier is shown in Fig. 9.

The bridge rectifier transforms the alternating current (AC) sinusoidal input voltage to dc voltage via a bridge diode (Fig. 10).

If a capacitor is not added into the rectifier circuit, the sinusoidal input voltage is converted to dc voltage as shown in Fig. 11.

Fig. 10 Input current bridge

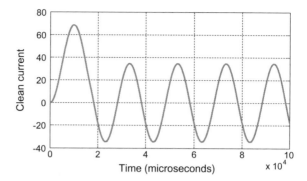

Fig. 11 Output voltage of bridge

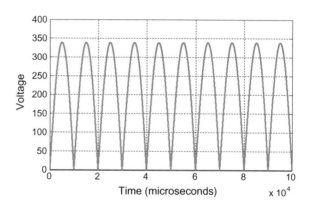

The output voltage of the bridge rectifier is calculated using Eq. 1,

$$Vdc = \frac{2}{T} \int\limits_{0}^{T/2} V_m \sin \omega t \, dt = \frac{2V_m}{\pi} = 0.636 V_m \qquad (1)$$

where V_m is the peak voltage.

If a capacitor is added but no load is connected to the output voltage, the output voltage will look like Fig. 12 (NL). The final output voltage without loading is a straight line. However, when it is loaded, the (WL) output voltage oscillates.

The input current is as shown in Fig. 13. The output current will have the same pattern as the output voltage, but it is much smaller.

If the input current of the bridge rectifier circuit is tested for total harmonics distortion, the THD value is 199.2% as shown in Fig. 14. This value is too far from the expected because this could feed back noise to the AC grid.

Fig. 12 Output voltage with load and no load bridge

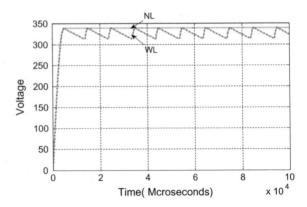

Fig. 13 Input current bridge type

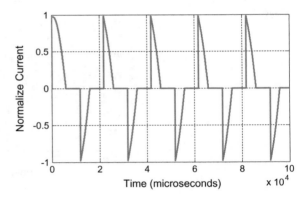

2.2.3 The Multi-level Inverter

The second element of the proposed battery charger is a multi-level inverter as shown in Fig. 15. Originally, the multi-level inverter will use dc supply voltage for the input and produce AC voltage at the output. However, for this particular battery charging application (as in battery charger), the inverter is fed by capacitor voltage from the bridge rectifier at the input and produces AC voltage at the output.

The output voltage of this multi-level is calculated as in Eq. (2).

$$V_o = \left(\frac{2}{T_O} \int_0^{T_O/2} V_S^2 dt \right)^{1/2} = V_s \tag{2}$$

The multi-level inverter produced an output as shown in Fig. 16. However, if the output is filtered, it produces a sinusoidal voltage.

The input current of this multi-level inverter is as shown in Fig. 17.

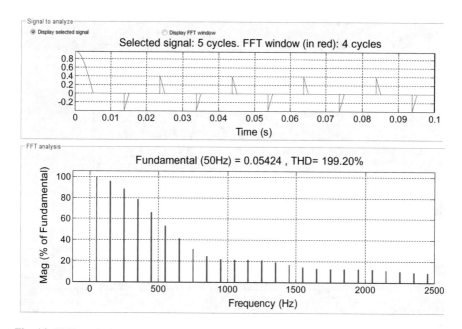

Fig. 14 THD analysis on bridge input current

Fig. 15 Multi-level inverter

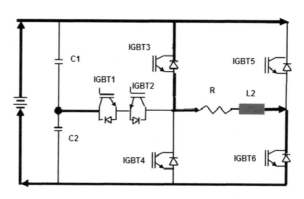

Fig. 16 Output voltage multi level type

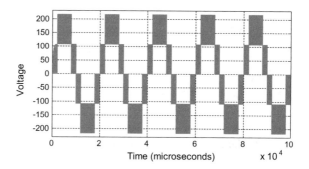

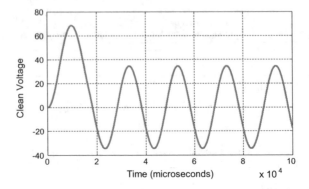

Fig. 17 Input current multi level type

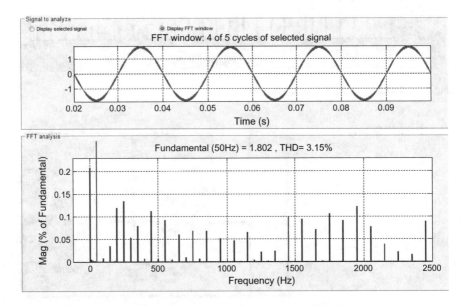

Fig. 18 THD analysis on multi level type

If the AC input current is tested for THD, the result will be shown in Fig. 18. The THD value is 3.15%, and it is acceptable.

2.2.4 The Buck Chopper

The third element of the proposed battery charger is a buck chopper as shown in Fig. 19. The purpose of the buck chopper is to regulate the output voltage level to

Fig. 19 Buck chopper

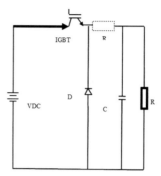

the desired voltage for battery charging. The IGBT is used to chop the input voltage to get the desired voltage at the output of the buck chopper. The inductor is used to lag the current, and the capacitor is used to lag the voltage.

The mathematical modelling for the buck chopper is explained as follows;

$$Vo(s) = \frac{V_{in}(s)}{S^2LC + RC(s) + 1} \tag{3}$$

For the PID controller software programming, the value of K_p, K_i and K_d are determined by, $K_p = 12.24$, $K_i = 24.5$, $K_d = 0$. The system is tested with these PID values, and the result is shown in Fig. 26.

3 Results and Discussion

Two simulation models were established to investigate the performance of each of the battery charger as shown in Figs. 20, 21 and 22.

Once the buck chopper is operating, the input and output of the buck chopper voltage are recorded as shown in Fig. 23. The voltage is reduced from 330 volts

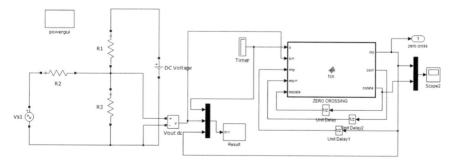

Fig. 20 Zero cross controller with offset circuit

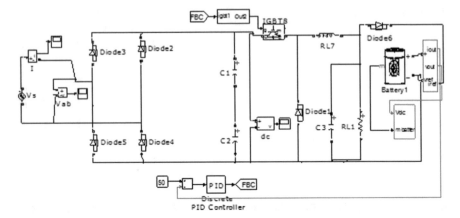

Fig. 21 Simulation of conventional bridge charger

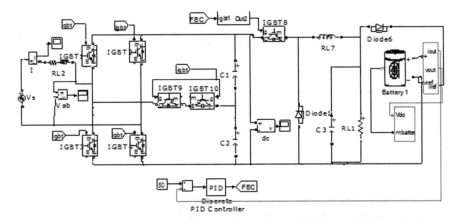

Fig. 22 Simulation model of multi-level rectifier

Fig. 23 Input and output voltage on buck chopper

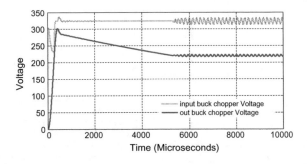

Fig. 24 Input voltage of
buck chopper fed from bridge
and multi-level type

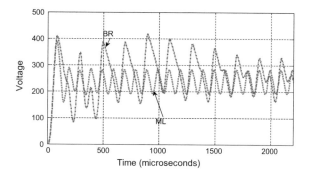

Fig. 25 Voltage analysis on
bridge and multi-level type

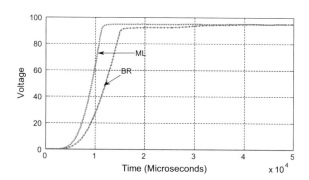

to 220 V. The output current has a similar pattern as the output voltage but with a smaller voltage level.

The output voltage of the multi-level rectifier is compared. The multi-level rectifier has low ripple voltage compared to the bridge rectifier as shown in Fig. 24. The output voltage of the multi-level and the bridge rectifier are compared. The multi-level rectifier has a lower ripple voltage compared to the bridge rectifier as shown in Fig. 24.

The output voltage of the buck chopper of the multi-level rectifier has a faster response compared to the bridge rectifier circuit is shown in Fig. 25.

The charging current of the multi-level has a faster response and has lesser ripple compared to the bridge rectifier as shown in Fig. 26.

Total harmonics distortion is tested using AC input current for both battery chargers, and the result is shown in Figs. 27 and 28. The multi-level rectifier has lower THD which is 4.68 compared to a bridge rectifier which is 49.63.

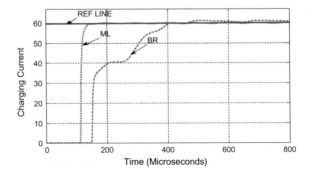

Fig. 26 Current analysis on bridge and multi-level type

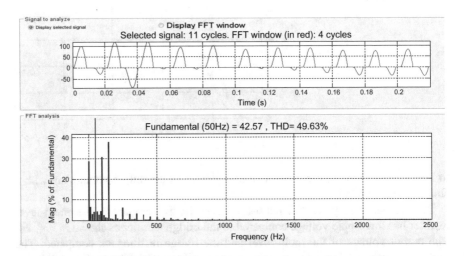

Fig. 27 THD analysis on bridge type

4 Conclusions

The proposed battery charger circuit using multi-level rectifier, buck Chopper and zero crossing circuit has better THD, fast response, less ripple and suitable to be used as a battery charger for EV home charging.

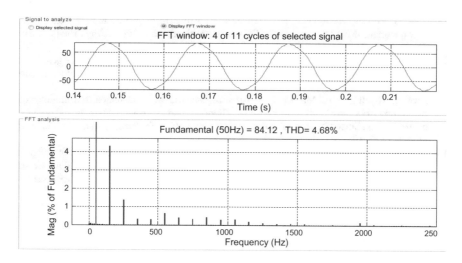

Fig. 28 THD analysis on multi level type

References

1. Gao, Y., Ehsani, M.: Design and control methodology of plug-in hybrid electric vehicles. IEEE Trans. Ind. Electron. **57**(2), 633–640 (2010)
2. Westbrook, M.H.: The Electric and Hybrid Electric Vehicle. SAE (2001)
3. Husain, I.: Electric and Hybrid Electric Vehicles, Design Fundamentals. CRC Press, Boca Raton
4. Oak Ridge National Laboratory: Advanced Brush Technology for DC Motors (2009). http://peemrc.ornl.gov/projects/emdc3.jpg
5. Ehsani, M., Gao, Y., Gay, S.E., Emadi, A.: Modern Electric, Hybrid Electric, and Fuel Cell Vehicles. CRC Press, Boca Raton (2005)
6. Emadi, A., Ehsani, M., Miller, J.M.: Vehicular electric power systems: land, sea, air, and space vehicles. Marcel Dekker, New York (2003)
7. Larminie, J., Lowry, J.: Electric Vehicle Technology Explained. Wiley, New York (2003)
8. Chan, C.C., Chau, K.T.: An overview of power electronics in electric vehicles. IEEE Trans. Ind. Electron. **44**(1), 3–13 (1997)
9. Arof, S., Jalil, J.A., Yaakop, N.M., Mawby, P.A., Arof, H.: Series motor four quadrants drive DC chopper. Part 1: overall. In: IEEE International Conference on Power Electronics (2014). https://doi.org/10.1109/pecon.2014.7062468
10. Arof, S., Khairulzaman, M., Jalil, A.K., Arof, H., Mawby, P.A.: Self tuning fuzzy logic controlling chopper operation of four quadrants drive DC chopper for low cost electric vehicle. In: 6th International Conference on Intelligent Systems, Modeling and Simulation, pp 24–40. IEEE Computer Society (2015). https://doi.org/10.1109/isms.2015.34
11. Arof, S., Khairulzaman, M., Jalil, A.K., Arof, H., Mawby, P.A.: Artificial intelligence controlling Chopper operation of four quadrants drive DC Chopper for low cost electric vehicle. Int. J. Simul. Sci. Technol. (2015). https://doi.org/10.5013/ijssst.a.16.04.03,2015.ijsst
12. Arof, S., Jalil, J.A., Kamaruddin, N.H., Yaakop, N.M., Mawby, P.A., Arof, H.: Series motor four quadrants drive DC chopper. Part 2: driving and reverse with direct current control. In: International Conference on Power Electronics, pp. 775–780 (2016). ISBN 978-1-5090-2547-3/16. https://doi.org/10.1109/pecon 2016.7951663

13. Arof, S., Hassan, H., Rosyidi, M., Mawby, P.A., Arof, H.: Implementation of Series motor four quadrants drive DC chopper for DC drive electric car and LRT. J. Appl. Environ. Biol. Sci. J. Appl. Environ. Biol. Sci. 7(3S), 73–82 (2017)
14. Arof, S., Noor,N.M, Elias, F., Mawby, P.A., Arof, H.: Investigation of chopper operation of series motor four quadrants DC chopper. J. Appl. Environ. Biol. Sci. J. Appl. Environ. Biol. Sci. 7(3S), 49–56 (2017)
15. Arof, S., Diyanah, N.H., Mawby, P.A., Arof, H.: Study on implementation of neural network controlling four quadrants direct current chopper: part 1: using single neural network controller with binary data output. In: Advanced Engineering for Processes and Technologies, pp. 37–57 (2019)
16. Arof, S., Diyanah, N.H., Yaakop, N.M., Mawby, P.A., Arof, H.: Processor in the loop for testing series motor four quadrants drive direct current chopper for series motor driven electric car: part 1: chopper operation modes testing. In: Advanced Engineering for Processes and Technologies, pp. 59–76 (2019)

A New Four Quadrants Drive Chopper for Separately Excited DC Motor in Low Cost Electric Vehicle

S. Arof, N. H. N. Diyanah, N. M. Noor, J. A. Jalil, P. A. Mawby and H. Arof

Abstract Four quadrants DC chopper systems are widely used in dc drive traction for electric vehicles. However, detail information on the design and method of operation for the systems were rarely addressed in the research literature. Accordingly, this study aimed to contribute on a new topology of a Four Quadrants Drive DC Chopper for separately excited dc motor. The chopper is designed to operate in five operation modes; driving, field weakening, generation, regenerative braking and resistive braking for the application of a low cost Electric Vehicle. The chopper modes of operation are further described and simulated using MATLAB/ SIMULINK. Results on chopper performance, i.e. switching power losses, ripple torque and current, voltage drop and output power, regenerative braking power and control were discussed. The proposed chopper operations have been verified through experimental setup and the chopper is observed to be capable of performing the expected operations.

Keywords DC drive · Separately excited machine · FQDC · DC chopper

S. Arof (✉) · N. H. N. Diyanah · N. M. Noor · J. A. Jalil
Electrical Electronic Automation Section, Universiti Kuala Lumpur, Malaysian Spanish Institute, Kulim Hi-Tech Park, 09000 Kulim, Kedah, Malaysia
e-mail: saharul@unikl.edu.my

N. H. N. Diyanah
e-mail: diyanahhisham94@gmail.com

N. M. Noor
e-mail: noramlee@unikl.edu.my

J. A. Jalil
e-mail: Julaida@unikl.edu.my

S. Arof · P. A. Mawby
University of Warwick School of Engineering, Coventry CV47AL, UK
e-mail: p.a.mawby@warwick.ac.uk

H. Arof
Engineering Department, Universiti Malaya, Jalan Universiti, 50603 Kuala Lumpur, Malaysia
e-mail: ahamzah@um.edu.my

© Springer Nature Switzerland AG 2019
M. H. Abu Bakar et al. (eds.), *Progress in Engineering Technology*,
Advanced Structured Materials 119, https://doi.org/10.1007/978-3-030-28505-0_10

1 Introduction

The emission of hydrocarbons not only pollutes the environment but also contributes to global warming, which melts the iceberg and increases the sea level. Using efficient Electric Vehicles (EV) and Hybrid Electric Vehicles (HEV) for transportation is one of the solutions to reducing global hydrocarbon emission. Unfortunately, the price of EV and HEV is expensive, making it unattainable for many people, especially those living in poor countries. Thus, there is a need for an efficient, compact drive system for EV and HEV that can reduce their cost and thus making them economical and affordable left [1–8] as an alternative.

It is well known that separately excited DC motors are easier to control and more stable in any mode of operation than series and shunt DC motors. To date, separately excited DC motors have been used in many prototypes or products for EV and HEV such as in the Peugeot 106, Citroen Saxo, GM EV and Lada [9, 10]. Research also has shown that separately-excited dc motor offers EV with longer distance traversed for a lower price [9] as depicted in Table 1. However, the last generation dc motors had disadvantages in their size, weight, performance and reliability. The separately excited DC motors can provide sufficient high starting torque and constant torque during operation. In the working region above base speed/nominal speed, DC motors still have higher torque compared to that of AC motors [11–14]. The motor can supply almost constant torque during their operation which is good for climbing a steep hill so that speed will not drop drastically. The motors also feature high electrical braking power due to less power loss [15, 16] and a very good speed and controllability for regenerative power. The regenerative power can be used to charge batteries or ultra-capacitors [17]. With the latest technology, DC motor manufacturers have developed better DC motors more

Table 1 Production of electric cars [2]

Manufacturer	Renault	Peugeot	Nissan
Model Name	Clio electric	106 electric	Hypermini
Driving type	Ac iduction	Separately excited	PM synch
Battery type	NiCd	NiCd	Li-ion
Max power O/P (kWh)	22	20	24
Voltage (V)	11.4	120	288
Battery energy capacity (kWh)	11.4	12	–
Top speed (km/h)	95	90	100
Claimed max range (km)	80	150	115
Charge time (h)	7	7–8	4
Price	$27,400	$27,000	$36,000

appropriate for EV/HEV applications. Such motors are equipped with higher power output, higher efficiency, smaller size, lighter weight, longer lasting carbon brush and commutator, lower operating voltage (less than 15 V) and easier to maintain structure (using modular concept construction) [10, 11]. The advanced brush technology for DC motors allows the motor to be used at low voltages [<50 V (35 V, 4500 A, 55 kW)] [11, 12] which results in lower power loss that guarantees longer travel distance.

In this paper, a novel four quadrants drive DC Chopper (FQDC) design shown in Fig. 1 is proposed to work with a separately excited DC motor. The system has resistive braking mode to achieve higher efficiency during braking. While the common H-bridge [15–17] shown in Fig. 2 utilizes at least two semiconductors at a time, the proposed FQDC uses only one semiconductor with the same operation. The proposed chopper design helps to improve power as well as to reduce ripple.

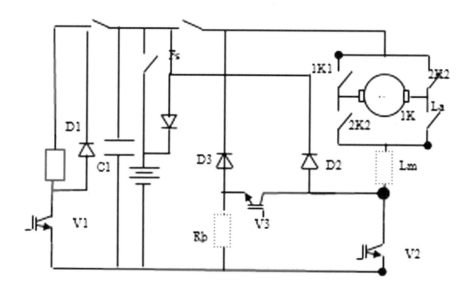

Fs= Main contactor
1K1 = Forward Contactor
2K2= Reverse Contactor
V1 = IGBT 1
V2= IGBT 2
V3 = IGBT 3

D2 = Armature free wheel diode
D3 = Bridging diode
Lm= Additional inductor
Rb = Braking resistor
D1 = Field free-wheel diode

Fig. 1 Proposed FQDC topology

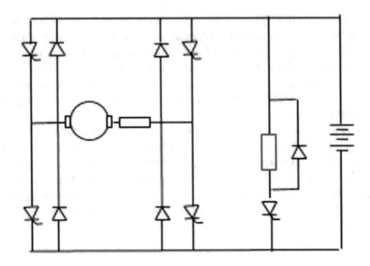

Fig. 2 Common H-bridge topology

2 Methodology

2.1 Four Quadrants Drive DC Chopper Operation Modes

The proposed FQDC is designed to work in five modes of operation; driving, field weakening, generator, regenerative braking and resistive braking. These modes of operation can be controlled by an AI controller such as expert system, fuzzy logic, self tuning fuzzy [2–4, 8], neural network as well as ANFIS. However, this paper focused only on the circuit operation of each modes of the proposed topology. Equations (1)–(4) are applicable for all chopper operation modes which represent the voltage, torque and current for the chopper. In those equations, B_{emf} is the back emf of the motor, K_b is the back emf constant and K_t is the torque motor constant. R_a and R_f are the motor coil resistances, ω is the angular speed and T_d is the motor torque. L_a is the armature inductance of the motor and L_m is an additional inductance connected in series with the armature windings to smoothen the motor armature current. V_t is the terminal voltage, L_f is the inductance of the field winding and R_b is the brake series resistor.

$$I_a = \frac{V_{batt} - I_a(R_a) - B_{emf}}{L_a} \tag{1}$$

$$I_f = \frac{V_{batt} - I_a\left(+R_f\right)}{L_f} \tag{2}$$

$$B_{emf} = K_v I_f \omega \qquad (3)$$

$$T_d = K_t I_f I_a = J\frac{d\omega}{dt} + B_w + T_L \qquad (4)$$

2.2 Driving Mode

The successes of driving mode rely on the capacitor and the charge that is stored. Field and armature coil of separately excited motor has different resistance values where the field has almost a ten times or higher resistance value than the armature winding. During starting up, the back emf is zero without back emf to limit the armature current; the current would be high at the armature. Unlike power supply at home, an EC Battery has only certain limited current that it could provide. Due to big difference of armature and field winding the current from the battery is most likely preferred to flow to the armature coil living very small amount entering the field winding. So the lack of current will be topped by the capacitor current. This action is only required at the start of the motor rotation because once it has already started the motor from the armature current will decrease as a result of BEMF. In this case the current from battery is sufficient to be segregated to the armature and field winding. After releasing the pulses, the armature/load current that is governed by the preset control factor α (for the field igbt) and β (for the armature igbt) flows through the chopper. The switching of main IGBT $V1$ and IGBT $V2$ are determined by the control factor of α and β related to the duty ratio. The current path for driving operation is shown in Fig. 3.

In driving mode, the control factor α is set to unity and the control factor β is varied. This is an armature control mode. IGBT $V2$ current will rise when the IGBT

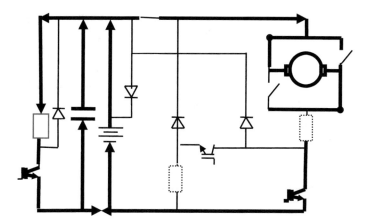

Fig. 3 FQDC in driving mode

V2 gate is fired on. The current rises gradually as its rate is limited by the circuit inductance. The values of armature and field currents when IGBT *V1* and *V2* are fired as in driving and field weakening mode can be described by Eqs. (5) and (6).

$$\left[-\left(\frac{R_f}{L_f}\right)\right][I_f]\left[\frac{1}{(L_f)}\right][V_{dc}K_{v1}] \tag{5}$$

$$\left[\frac{-R_a}{L_a}\right][I_a]\left[\frac{1}{L_a}-\frac{1}{L_a}\right]\left[\begin{array}{c}V_{dc}K_{v2}\\ B_{emf}\end{array}\right] \tag{6}$$

2.3 Field Weakening Mode

When the field weakening is needed, the control factor β of IGBT *V2* is set to almost unity. The control factor α of IGBT *V1* is gradually decreased. This is the field control mode and the current path is shown in Fig. 4.

Reducing the current distribution at field winding decreases the back emf voltage as described by Eq. (1). The value of the current in the field coil is indirectly proportional to the value of α. Meanwhile, the armature current I_a rises as the IGBT *V1* duty ratio is decreased in proportion with the reduction in field current. The motor speed will increase due to the inflow of the armature current. This leads to an increase in the motor torque and speed. The mathematical equations used in the driving mode are also applicable to the field weakening mode. Only the control strategy differs between the two modes of operation.

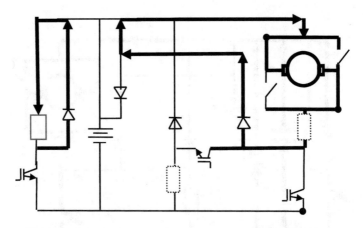

Fig. 4 FQDC in field weakening mode

2.4 Generator Mode

The kinetic energy that is available when the vehicle is going downhill can be used to generate electrical power to charge the batteries. The generated power is collected through the armature winding. The current has to flow through the field winding to establish the magnetic field. The IGBT *V1* is fired on to generate the voltage and current at the armature winding. The flow of current in the field and armature winding is shown in Fig. 5.

When contactor *2K2* is energized the armature current becomes negative due to the change in current direction. The generator mode reduces the motor speed since a counter torque motion is generated. In the generator mode, the field current and armature current are given by Eqs. (7) and (8).

$$\left[-\left(\frac{R_f}{L_f}\right)\right][I_f]\left[\frac{1}{(L_f)}\right][V_{dc}K_{v1}] \tag{7}$$

$$\left[\frac{-(R_{bh}+R_{batt}+R_a)}{L_a}\right][I_a]\left[\frac{1}{L_a}\right][B_{emf}] \tag{8}$$

2.5 Regenerative Braking Mode

The braking operation starts with the firing of IGBT *V1* until a minimum motor current is reached. As soon as IGBT *V1* is fired, current flows from the battery via IGBT *V1*. This guarantees a quick build-up of the motor voltage and currents via

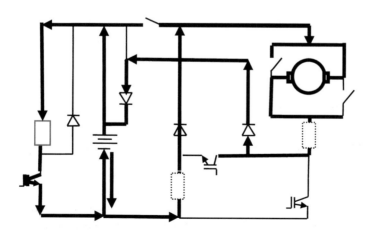

Fig. 5 FQDC in generator mode

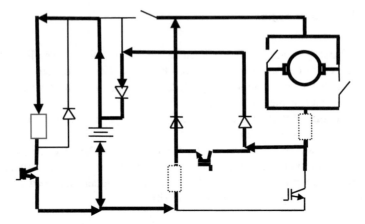

Fig. 6 FQDC in regenerative braking mode

armature terminal. The path of both excitation and load currents in this mode of operation is shown in Fig. 6.

Every time IGBT *V3* is triggered, it creates a short circuit path for the armature current to flow. Instead of going to the batteries to generate voltage, the current goes back to the armature coil which is short circuited. Due to the short circuit, a high current is induced. This high current creates a counter torque action against the motion of the motor which leads to braking or slowing down of the motor speed. The braking action reduces the kinetic energy of the vehicle and consequently the rotational speed of the motor. The armature and field current during regenerative braking can be represented by Eqs. (9) and (10).

$$\left[\frac{-(R_f)}{L_f}\right][I_f]\left[\frac{1}{L_f}\right]\left[V_{dc}K_{v1}\Big/\frac{B_{emf}}{(1-K_{v3})}\right] \tag{9}$$

$$\left[\frac{-(R_a+R_{batt})}{(L_a)}\Big/\frac{-(R_a+R_f)}{(L_a+L_f)}\right][I_a]\left[\frac{1}{(L_a)}\right]\left[\frac{B_{emf}}{(1-K_{v3})}\right] \tag{10}$$

2.6 Resistive Braking

The IGBT *V1* is fired to build up the needed generated voltage. These actions are illustrated as in Fig. 7.

The resistive braking mode starts when IGBT *V2* is fired. The duty ratio of IGBT *V2*, namely β, is varied according to the speed of the motor. IGBT *V1* is fired to keep the motor current constant. The generated voltage reduces proportionally with the decreasing vehicle speed. The firing of IGBT *V2* guarantees the motor current to commutate into the braking resistor R_b. The flow of the current can be regulated by

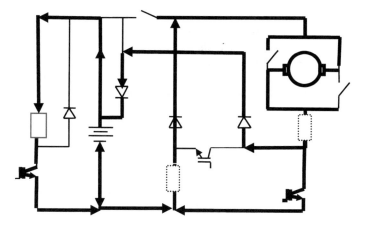

Fig. 7 FQDC in resistive braking mode

controlling the duty ratio of IGBT *V2*. The armature and field current during resistive braking can be represented by Eqs. (11) and (12) respectively.

$$\left[\frac{-\left(R_f\right)}{L_f}\right][I_f]\left[\frac{1}{L_f}\right]\left[V_{dc}K_{drv}\Big/\frac{B_{emf}}{\left(1-K_{v3}\right)}\right] \tag{11}$$

$$\left[\frac{-\left(R_a+R_{bh}+R_{batt}\right)}{\left(L_a\right)}\Big/\frac{-\left(R_a+R_{bh}+R_f\right)}{\left(L_a+L_f\right)}\Big/\frac{-\left(R_a+R_{bh}\right)}{\left(L_a\right)}\right][I_a]\left[\frac{1}{\left(L_a\right)}\right]\left[\frac{B_{emf}}{\left(1-K_{rgb}\right)}\Big/B_{emf}K_{v2}\right]$$
$$\tag{12}$$

This type of braking mode can work independently or in combination with regenerative braking as shown in Fig. 8.

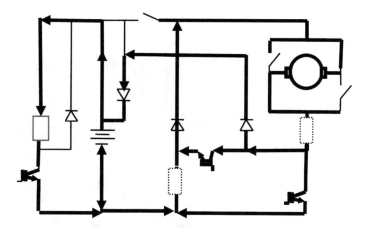

Fig. 8 Regenerative and resistive braking mode

If it works in combination the produce braking torque would be higher than the regenerative braking suitable for emergency brake. To work in combination the braking resistor value is set close to the regenerative braking short circuit resistance. The braking power of standalone resistive braking mode is much related to the value of braking resistor. The higher the braking resistor the lower the current flow through the braking resistor and the lower the braking torque produced.

2.7   Experimental Setup

In order to test the FQDC in all modes, an experimental set-up as shown in Figs. 9 and 10 were prepared. A 650 W separately excited dc motor shaft is coupled with another motor with inertia load (flywheel). Experimental results were captured using a data acquisition software. The experimental values of motor speed, armature current and field current were transferred to FQDC model in MATLAB/ SIMULINK to validate the simulation model.

2.8   Results and Discussion

The Experimental and simulation results of the motor speed, armature current and field current for each modes of operation were shown in Figs. 11, 12, 13, 14 and 15. At the beginning of the driving mode, motor speed increases until it reaches the saturation level as shown in Fig. 11. Then the field weakening mode shown in Fig. 12 was initiated causing the motor speed to increase due to the rise in torque. In this mode, the armature current increases while the field current decreases.

Fig. 9 FQDC for separately excited motor

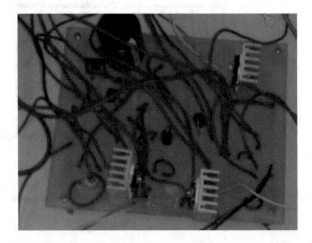

Fig. 10 Experimental set-up for the FQDC

Fig. 11 Driving mode

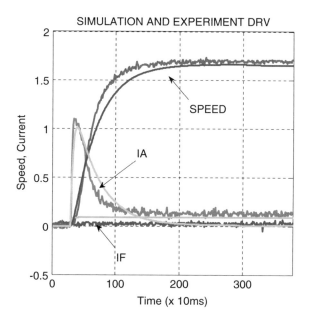

In the generator mode shown in Fig. 13, the field winding was charged with voltage to establish the magnetic field so that the armature coil could produce the generated voltage. In the graph the generated voltage shown has been normalized. During regenerative braking shown in Fig. 14, IGBT *V3* is turned on causing a short circuit path at the armature. The armature current rose in the opposite direction resulting in the developed torque to oppose the motor motion.

This caused the braking action. In the resistive braking mode shown in Fig. 15 the generated voltage was dissipated through the braking resistor. It is thus observed that the proposed chopper design behaved like a buck converter in all

Fig. 12 Field weakening
mode

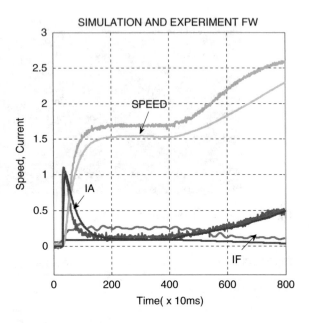

Fig. 13 Generator mode

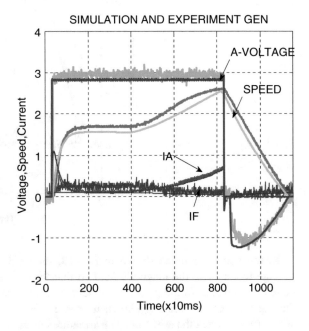

operation modes except for regenerative braking. In regenerative braking mode, the chopper was observed to behave like a buck-boost converter.

The proposed FQDC offers several favorable performances such as lower power switching losses, smoother torque and current ripple, less voltage drop and higher

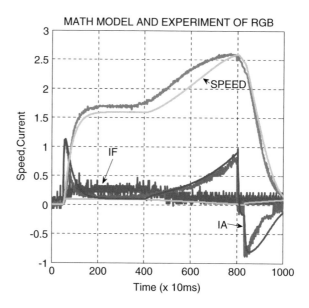

Fig. 14 Regenerative braking mode

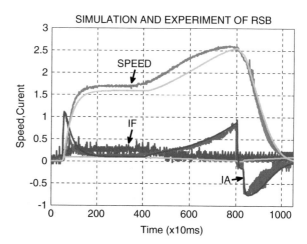

Fig. 15 Resistive braking mode

output power, as well as better braking performance as compared to the conventional H-bridge chopper. Every switching power electronics contributes to power switching losses. Losses during switch off are far greater than during switch on. The proposed chopper has single IGBT and single diode switching on and off during its operation. Meanwhile, the H-bridge has two power semiconductors and two diodes switching on and off during its operation. Thus, the switching power losses in H-bridge would be doubled. Figures 16, 17, 18 and 19 show the turn on and turns off power losses in IGBT and diodes respectively.

Fig. 16 IGBT turn on power
losses

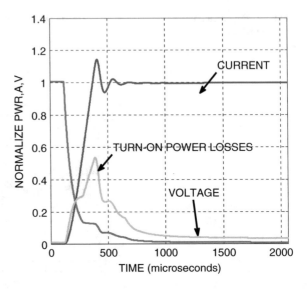

Fig. 17 IGBT turn off power
losses

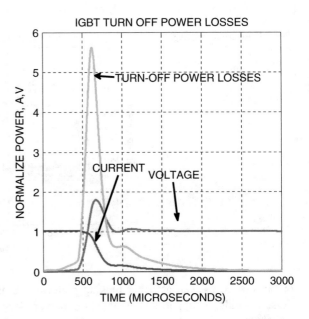

The proposed chopper also offers smoother armature current and torque shown
in Figs. 20 and 21, as compared to H-bridge due to a single IGBT firing. It is
difficult to get the IGBT in H-Bridge fires at the same time. The proposed chopper
used contactor so that the voltage stored in inductor winding will be easily
discharged.

Fig. 18 Diode turn on power losses

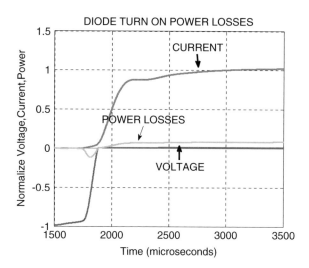

Fig. 19 Diode turn off power losses

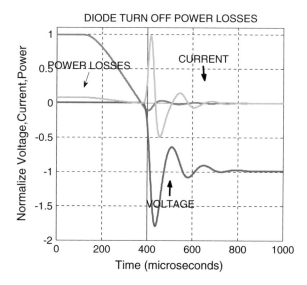

Accordingly, voltage at the inductor winding during IGBT turn off can easily be removed in this chopper circuit as compared to voltage in the H-bridge. Voltage stored in H-bridge has to be returned to the battery. When the stored voltage is lower than the battery voltage, the clamping diode will switch off thus blocked the stored voltage avoiding it to return to its origin. When the IGBT is switch on again, the residual voltage stored in the circuit causes ripple in armature current. The influence of low ripple has current results in low ripple torque.

Every semiconductor has voltage drop during turn on. This is called the voltage collector emitter saturation. While the proposed chopper running on single IGBT,

Fig. 20 Ripple current

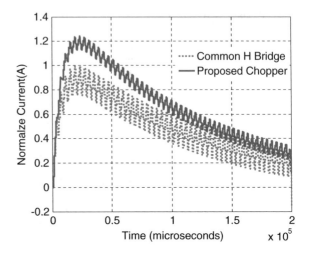

Fig. 21 Ripple torque

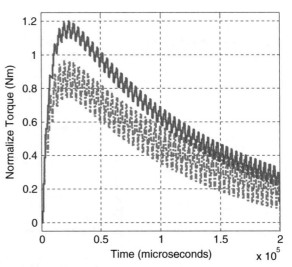

H-bridge is running on two IGBT resulted in a double voltage drop. Since the chopper has less voltage drop as compared to the one in H-bridge, therefore its armature current is observed to be higher than in H-bridge. As higher current produces more torque and higher speed, thus the chopper output power is observed to be higher. The current and output power is depicted in Figs. 22 and 23 respectively.

As the battery state of charge is almost 100%, the voltage difference between the charging voltage and battery terminal voltage is getting lower and resulting in a lower charging current. However, braking is required at this moment. Since the battery current is low, thus armature current is also low. Accordingly the braking torque produced is also low. In order to achieve the required braking torque, IGBT *V3* is fired. The IGBT *V3* caused a rise in armature current due to the effect of

Fig. 22 Armature current

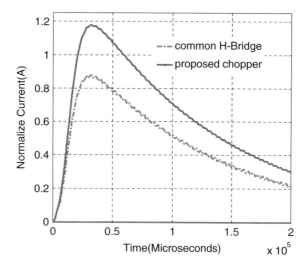

Fig. 23 Braking effect at high SOC

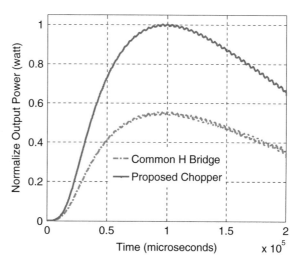

armature voltage short circuited. Consequently, the braking torque is increased as shown in Fig. 24. In the case of higher braking torque is needed, i.e. an emergency brake, IGBT *V3* can be further controlled to produce higher braking torque as shown in Fig. 25 thus the EV speed will decelerate faster. Since the braking torque is influenced by both field and armature currents, therefore any changes in vehicle speed will affect the field current. Subsequent change in vehicle speed leads to change in braking torque. Since the braking torque produced is always nonlinear, thus a proper control of IGBT *V3* provides better control in the braking torque, resulting in linear braking effect as shown in Fig. 25.

Resistive braking mode offers another type of electrical braking. In the case of a higher braking action is required, a lower braking value of braking resistor is

Fig. 24 Higher braking effect in regenerative braking

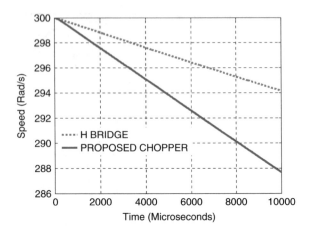

Fig. 25 Linear braking efect in regenerative braking

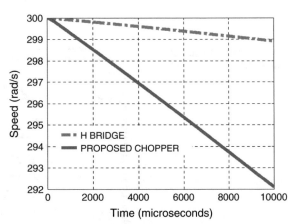

Fig. 26 Resistive braking effect

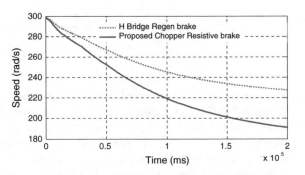

chosen. As lower braking resistor causes a rise in armature current, hence higher braking torque can be generated resulting a drop in motor speed as shown in Fig. 26. Resistive braking mode is able to extend the electrical braking before mechanical brake takes over.

3 Conclusions

The proposed four quadrants drive DC chopper design has been simulated and experimented to drive a separately excited DC motor for five modes of operation. The experimental results revealed that chopper is capable to operate in all the five modes. However, the FQDC performances such as lower power switching losses, smoother torque and current ripple, less voltage drop and higher output power, as well as better braking performance as compared to the conventional H-bridge chopper were simulated using Matlab/Simulink mathematical model. In summary, the FQDC has great potential to be utilized in EV application due to simple and excellence controllability thus made it suitable for applications in this cost effective EV solution.

References

1. Arof, S., Jalil, J.A., Yaakop, N.M., Mawby, P.A., Arof, H.: Series motor four quadrants drive DC chopper. Part 1: overall. In: IEEE International Conference on Power Electronics (2014). https://doi.org/10.1109/pecon.2014.7062468
2. Arof, S., Khairulzaman, M., Jalil, A.K., Arof, H., Mawby, P.A.: Self tuning fuzzy logic controlling chopper operation of four quadrants drive DC chopper for low cost electric vehicle. In: 6th International Conference on Intelligent Systems, Modeling and Simulation, pp 24–40. IEEE Computer Society (2015). https://doi.org/10.1109/isms.2015.34
3. Arof, S., Khairulzaman, M., Jalil, A.K., Arof, H., Mawby, P.A.: Artificial intelligence controlling Chopper operation of four quadrants drive DC Chopper for low cost electric vehicle. Int. J. Simul. Sci. Technol. (2015). https://doi.org/10.5013/ijssst.a.16.04.03,2015.ijsst
4. Arof, S., Jalil, J.A., Kamaruddin, N.H., Yaakop, N.M., Mawby, P.A., Arof, H.: Series motor four quadrants drive DC chopper. Part 2: driving and reverse with direct current control. In: International Conference on Power Electronics, pp. 775–780 (2016). ISBN 978-1-5090-2547-3/16. https://doi.org/10.1109/pecon 2016.7951663
5. Arof, S., Hassan, H., Rosyidi, M., Mawby, P.A., Arof, H.: Implementation of Series motor four quadrants drive DC chopper for DC drive electric car and LRT. J. Appl. Environ. Biol. Sci. J. Appl. Environ. Biol. Sci. 7(3S), 73–82 (2017)
6. Arof, S., Noor,N.M, Elias, F., Mawby, P.A., Arof, H.: Investigation of chopper operation of series motor four quadrants DC chopper. J. Appl. Environ. Biol. Sci. J. Appl. Environ. Biol. Sci. 7(3S), 49–56 (2017)
7. Arof, S., Diyanah, N.H., Mawby, P.A., Arof, H.: Study on implementation of neural network controlling four quadrants direct current chopper: part 1: using single neural network controller with binary data output. In: Advanced Engineering for Processes and Technologies, pp. 37–57 (2019)
8. Processor in the loop for testing series motor four quadrants drive direct current chopper for series motor driven electric car: part 1: chopper operation modes testing. In: Advanced Engineering for Processes and Technologies, pp. 59–76 (2019). https://doi.org/10.1007/978-3-030-05621-6_5
9. Westbrook, M.H.: The Electric and Hybrid Electric Vehicle. SAE (2001)
10. Husain, I.: Electric and Hybrid Electric Vehicles, Design Fundamentals. CRC Press, Boca Raton
11. Oak Ridge National Laboratory: Advanced Brush Technology for DC Motors (2009). http://peemrc.ornl.gov/projects/emdc3.jpg

12. Oak Ridge National Laboratory: Soft-commutated direct current (dc) motor (2009). Available: www.ornl.gov/etd/peemrc
13. Heinrich, Walter Rentsch, Herbert Dr.-Ing., ABB Industry: guide to variable speed drives. Technical Guide No. 41180 D-68619 LAMPERTHEIM, Germany, 3ADW 000 059 R0201 REV B (02.01), DCS 400/DCS 500/DCS 600: ABB (2003)
14. Bansal, R.C.: . Electric vehicle. In: Handbook of Automotive Power Electronics and Motor Drives. Taylor & Francise Group, CRC Press, Boca Raton
15. Rashid, M.H.: Power Electronics, Circuits, Devices and Applications, 3rd ed. Prentice Hall, Upper Saddle River
16. Senthil Kumar, N., Sasasivam, V., Muruganandam, M.: A low cost four-quadrant chopper—fed embedded DC drive using fuzzy controller. J. Electr. Power Comp. Syst. **35**, 907–920
17. Joshi, D., Bansal, R.C.: Performance evaluation of multiquadrant DC drive using fuzzy-genetic approach. J. Electr. Syst. (JES) **5**(4) (2009)

Genetics Algorithm for Setting Up Look Up Table in Parallel Mode of Series Motor Four Quadrants Drive DC Chopper

S. Arof, N. H. N. Diyanah, N. M. N. Noor, M. Rosyidi, M. S. Said, A. K. Muhd Khairulzaman, P. A. Mawby and H. Arof

Abstract This paper presents the establishment of look-up table (LUT) for speed versus field current. LUT is necessary as an input reference for the close loop control of field current. The LUT was developed with the assistance of a genetic algorithm (GA). GA was used for optimization of the field current to maintain the motor torque for DC series motor running in parallel mode. Other than that, the LUT was employed when the FQDC was run in parallel mode for the purpose of climbing up a steep hill or slope. The simulation results using MATLAB/Simulink showed that the DC series motor with the assistance of LUT could overcome the drawbacks of DC series motor as speed decreases drastically when climbing a steep

S. Arof (✉) · N. H. N. Diyanah · N. M. N. Noor · M. Rosyidi ·
M. S. Said · A. K. Muhd Khairulzaman
Electrical Electronic Automation Section, Universiti Kuala Lumpur,
Malaysian Spanish Institute, Kulim Hi-Tech Park, 09000 Kulim,
Kedah, Malaysia
e-mail: saharul@unikl.edu.my

N. H. N. Diyanah
e-mail: diyanahhisham94@gmail.com

N. M. N. Noor
e-mail: noramlee@unikl.edu.my

M. Rosyidi
e-mail: mrosyidi@unikl.edu.my

M. S. Said
e-mail: msazali@unikl.edu.my

A. K. Muhd Khairulzaman
e-mail: khairulzaman@unikl.edu.my

S. Arof · P. A. Mawby
University of Warwick School of Engineering, Coventry CV47AL, UK
e-mail: p.a.mawby@warwick.ac.uk

H. Arof
Engineering Department, Universiti Malaya, Jalan Universiti,
50603 Kuala Lumpur, Malaysia
e-mail: ahamzah@um.edu.my

© Springer Nature Switzerland AG 2019
M. H. Abu Bakar et al. (eds.), *Progress in Engineering Technology*,
Advanced Structured Materials 119, https://doi.org/10.1007/978-3-030-28505-0_11

hill. In conclusion, the GA had successfully determined the best optimum field current to produce the highest torque to overcome load effect when tested using the proposed Four-Quadrant DC Chopper.

Keywords DC drive · EV and HEV · Series motor · Four-quadrant chopper · Parallel mode · GA · LUT

1 Introduction

In future, the electric motor propulsion system (electric vehicle) will replace the internal combustion system (mechanical combustion engine). This is not only due to its zero emission but also because of its higher efficiency [1–3]. The advancement of supercapacitor as an alternative to batteries is the catalyst. In fact, at present, the supercapacitor, which is as big as 12,000 F, is already available in the market, and it has been expected that the development of a bigger storage capacity is in progress in research laboratories worldwide.

In 2003 US based Research Center Oak Research National Laboratory (ORNL) developed a 55 kW highly efficient brushed motor (92%) that operated at 13 V [4]. Moreover, it has been expected that it is currently producing lower than 5 V operating voltage to suit the electric car application.

2 Methodology

2.1 A Proposed Design of Four-Quadrants Drive DC Chopper

A DC drive with a separately excited DC motor has once been the best performance prototype electric car for low cost and maximum distance traversed. A separately excited DC motor normally uses two sets of batteries which are for armature and field winding cause the vehicle is heavy which affect the overall performance. DC series motor on the other hand the highest starting torque, less weight, simpler drive control can run with a single battery source, but it has poor performance when loaded and tends to overrun while running without load [4]. To overcome this problem, a new proposed four quadrants DC chopper (FQDC) application of electric car (EC) is proposed [5–12]. The proposed chopper has several modes of operation such as driving, field weakening, generator, parallel, regenerative braking and finally resistive braking. Parallel mode is used to overcome the load effect such as if the EC is climbing a steep hill. The common H-bridge chopper is unable to perform field weakening, resistive braking, and reverse rotation [5]. In this paper, a novel 4-quadrant DC chopper topology as in Fig. 1 is discussed. In fact, this novel chopper had been proven to solve inability of the common H-bridge chopper. This

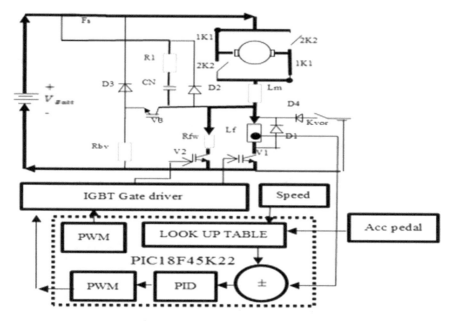

Fig. 1 Proposed four quadrants DC chopper

chopper also had exhibited the ability to perform several modes of operations, such as driving, field weakening, regenerative braking, resistive braking, generator, as well as reverse and parallel mode driving. However, this paper only focused on that related to parallel mode. This new design had been proven to solve the common problem of DC series motor, which is the linear decrease of speed when loaded [5].

2.2   The Proposed Four-Quadrant DC Chopper Design

The following general equations describe the voltage and current for the chopper. Equations (1)–(4) are general equations that are applicable to all chopper operation modes. B_{emf} is the back emf of the motor, V_a and V_f are the armature and field voltages, K_b is the back emf constant, K_t is the torque motor constant, I_f is the field current, R_a and R_f are the motor coil resistances, I_a is the armature current, ω is the angular speed, and T_d is the motor torque.

$$I_a = \frac{V_{batt} - I_a(R_a + R_f) - B_{emf}}{L_a + L_f} \tag{1}$$

$$B_{emf} = K_v I_f \omega \tag{2}$$

$$T_d = K_t I_f I_a \tag{3}$$

$$T_d = J\frac{d\omega}{dt} + B_w + T_L \tag{4}$$

2.3 DC Series Motor and Control Strategy During Parallel Mode

A DC series motor has the advantage of high starting torque. However, there is one common weakness of DC series motor, which is the drastic drop in motor speed when loaded. Besides, it is common for any electrical motor to have a decrease in speed when loaded, but for series, it is obviously a disadvantage to such motor. Hence, in order to solve this problem, this paper had proposed four-quadrant choppers that allowed the series motor to operate in parallel mode. In this mode, the field current is separately controlled by the armature current, where it can be changed or maintained. Furthermore, when loaded, the speed of the motor would decrease. Decrease in speed would eventually result in reduced B_{emf} see Eq. (1). If B_{emf} is reduced, the armature current will rise see Eq. (2). Since the field current is controlled separately and can be maintained, the increase in armature current will raise the torque see Eq. (3). This new torque should be able to overcome the loading effect. Thus overcome the decrease in speed while climbing a steep hill. During parallel mode, the armature, and the field current can be represented by the following equations:

$$\frac{d}{dt}\left[\dot{I}_f\right] = \left[-\left(\frac{R_f}{L_f}\right)\right][I_f]\left[\frac{1}{L_f}\right][V_{ext}K_{drv}] \tag{5}$$

$$\frac{d}{dt}\left[\dot{I}_a\right] = \left[-\left(\frac{R_a}{L_a}\right)\right][I_a]\left[\frac{1}{L_a}\right]\left[\frac{V_{dc}}{E_g}\right] \tag{6}$$

In this parallel mode, the field and the armature can be separately controlled. Hence, higher torque can be produced if the armature current is set at a high value. Field current, on the other hand, is adjusted to have lower B_{emf}, but enough torque to overcome load.

2.4 The Control Strategy in Parallel Mode

In order to obtain a high torque, the armature current is set to maximum. This can be achieved by setting the PWM value of the IGBT V2 to the maximum. Besides, field current is adjusted by FIRING IGBT V1 to the value where it produces low B_{emf}. Nonetheless, this does not always mean that lower field current is better. In

Fig. 2 Variety of fields current produces different torque

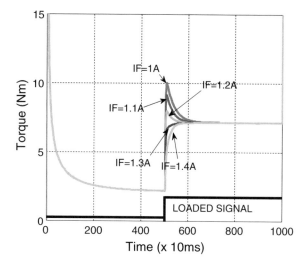

Fig. 3 Variety of fields current produces different torque result

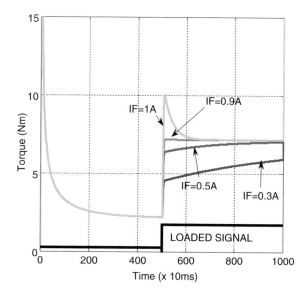

addition, to some extent, low in field current does not only result in lower B_{emf}, but lower in torque too. This is because; to some extent, lower field current results in lower magnetic field and flux. These two aspects are very important elements in producing the motor torque. Figure 2 shows the motor torque as a result of changes in field current. As illustrated in Fig. 2 and 3, the maximum field current to achieve the maximum torque is 1A.

Furthermore, Figs. 4 and 5 display the changes in motor speed with respect to changes in field current when loaded.

Fig. 4 Variety of fields
different current speed

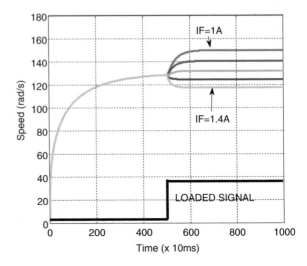

Fig. 5 Variety of fields
current different speed result

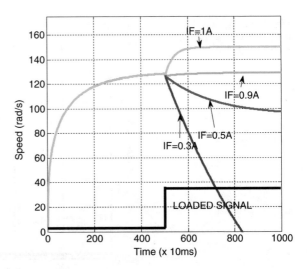

Hence, in order to overcome overfeeding of field current, a lookup table (LUT) as in Table 1 and Fig. 6 had been employed. In this LUT, the best optimum value of field current associated with the motor speed of vehicle and accelerator pedal had been set. However, in a real application, the value is always inclusive of a safety value to avoid overshoot. Thus, the LUT functioned as the library where the optimized reference field current value is stored. As such, a genetic algorithm or gradient can be used to optimize the required field current.

Table 1 Look up table

Speed	Field current
0–50 rqd/s	12–10
50–100	10–8
100–150	8–7

Fig. 6 Look up table

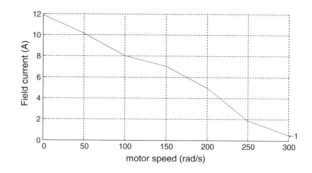

2.5 Genetic Algorithm

The genetic algorithm (GA) is based on natural genetics, which consists of a number of chromosomes in a population of individuals [13]. GAs are used as optimization and search algorithms, which apply the same concepts as natural genetics, mimicking evolution in nature [14]. The application of GA can also be used for setting up an LUT, for example, switching time LUT for the application of two-level inverter and chamfer distance LUT for image processing application [13–15].

The GA was employed to determine the optimum point of the field current to generate lower B_{emf}, and at the same time, to identify the optimum field current for maximum torque to overcome the load effect. The general LUT setup using GA operation is prescribed as a block diagram in Fig. 7.

The motor torque and speed are the two important signals for the GA controller. The output of the controller relies on these two values. Besides, the speed of the DC motor value is the input for GA operator. With respect to the speed, an equation was used to identify the region of the best field current where the optimum torque was determined.

The General equation is

$$Range_{field} = \frac{V_{dc}}{Speed} \times C \tag{7}$$

V_{dc} is the voltage supply, *Speed* is the spindle speed and C is the constant.

Besides, a more precise equation can be obtained from the fitness function equation. The GA controller and its control algorithm had been applied to refine this

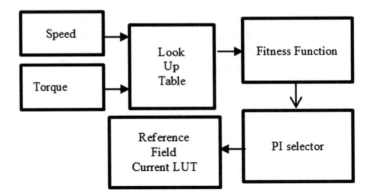

Fig. 7 Block diagram for parallel mode

region. When GA algorithm was activated, it first created randomly six chromosomes with six genes in each chromosome. The six chromosomes represented the six field current genes. The chromosomes were in decimal form, while the genes were in binary form. Each of these genes and chromosomes had been stored in memory for the further process. On the other hand, this field current represented falls. Each chromosome, which is in decimal number, was tested with fitness function to examine the strength of each chromosome. The fitness function equation was extracted from Eqs. (1) to (5), while those in steady state were represented by the following equations:

$$T_d = \frac{V_{batt} - K_v I_f w}{R_a} \times K_t I_f \tag{8}$$

In addition, this equation had been applied to test each chromosome. The actual output of this equation was the motor torque. Other than that, the fitness function test of individual field current was carried out to determine the best optimum torque. Moreover, weak chromosomes resulted in a lower value of output torque, while strong chromosomes resulted in higher output torque. Meanwhile, genes with a lower value of fitness function were eliminated, mutated or had the genes to change over with other chromosomes with a high fitness function. The remaining genes, which produced high torque, had their chromosomes exchanged, resulting in new genes and new chromosomes in a new population. Furthermore, the process of choosing the good chromosomes was done via the roulette wheel method. The mutation process reflected when an individual chromosome gene, which was in binary form, changed its bits form. Thus resulting in new genes and new chromosomes. From time to time, the genes of all the chromosomes were generated to be single uniform chromosomes with similar values of the genes. The process was repeated depending on the set of a number of repetition preset or the number of iteration. If the number of iteration elapsed, the GA algorithm would stop and wait for the next cycle to be reactivated. Changes in speed reactivated the GA controller. When the GA had finally stopped, it would have already found the best chromosomes and the system would have finally identified the best field current value with respect to the motor speed. The flow of operation is described in the flow chart as in Fig. 8.

3   Simulation Model and Results

The simulation model is the faster, cost and time effective, as well as a safer method of testing any system. A simulation model of Four-quadrant DC chopper, PID controller, GA controller was set up by using MATLAB/Simulink software. Moreover, two models were tested; a small kilowatt DC series motor and a DC Drive Electric Car. The models were developed by solving the linear Differential equation of DC series motor and had been represented in physical-based modelling using MATLAB/Simulink model form. A DC series motor with the 0.6 KW motor model was first simulated to determine the load, and it was controlled by four-quadrant chopper with PID, and the reference field current was provided by LUT. The results are shown in Figs. 9, 10, 11, 12 and 13.

Figure 9 is the first generation of the genes tested with a fitness function. Three genes are projected the maximum torque value. The other genes had been transformed into these genes and thus, could not be seen as they overlapped each other. On the other hand, Fig. 10 shows only two genes and in Fig. 11 refers to the ninth generation, and only one single gene was observed due to overlap. Furthermore, Fig. 12 shows in the seventh generation of the genes the best value of torque had been found. In Fig. 13 the LUT for torque versus speed is shown. Figure 14 shows that the torque output was generated when the LUT was tested with FQDC and series motor in parallel mode. It had been obvious that the parallel mode using LUT produced a higher torque. Thus, LUT was examined with another test, which was the DC EV test, as shown in Fig. 15. The LUT data scaled up due to higher DC series motor as Kilowatt was used. The DC drive EV test model consisted of FQDC chopper, vehicle dynamics, IGBT firing PD controller and LUT.

Furthermore, the model consisted of the chopper, vehicle dynamics, IGBT firing controller, and GA controller model. Meanwhile, in the vehicle dynamics model, the earth profile was set so that the test involved hill climbing, as shown in Fig. 15. Moreover, for comparison, the series motor configuration was also simulated, and the result is depicted in Fig. 16.

As a result, the parallel mode with the assistance of LUT displayed better performance. The speed drop was successfully slowed down, but without the parallel mode, the EV experienced a decrease in speed.

Fig. 8 Flowchart of GA operation

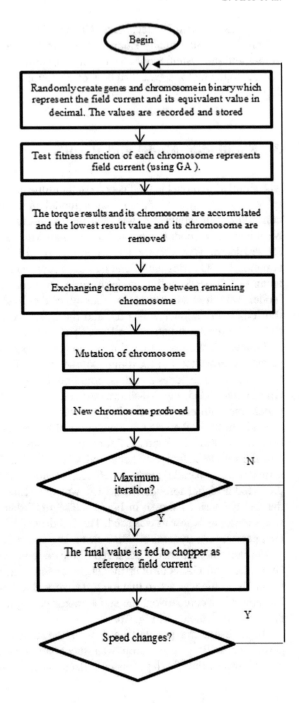

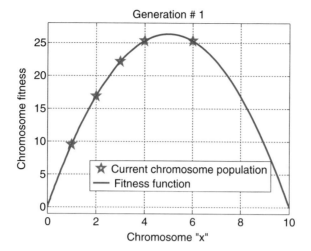

Fig. 9 Chromosome in first generation

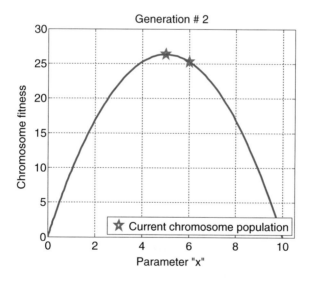

Fig. 10 Chromosome in second generation

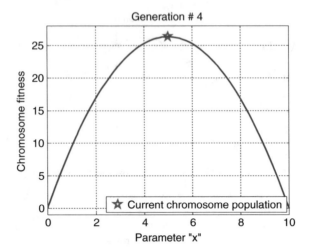

Fig. 11 Chromosome in fourth generation

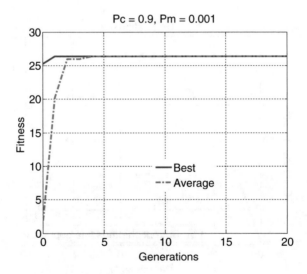

Fig. 12 Best and average results at each generation

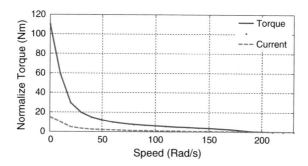

Fig. 13 .

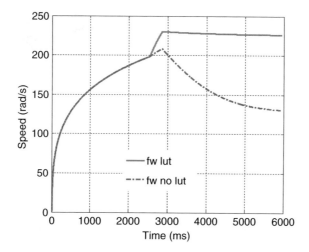

Fig. 14 Speed tested with LUT

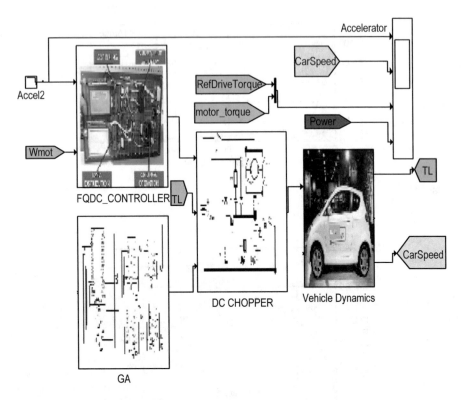

Fig. 15 MATLAB/Simulink model

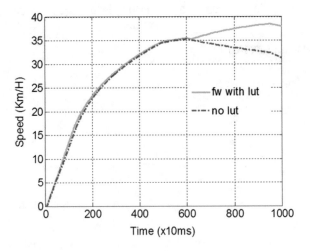

Fig. 16 LUT test with EV results

4 Conclusions

The DC drive series motor displayed a high potential to be utilized in EV. This had been due to its simple design, low cost, and excellent controllability. The GA had successfully formed the LUT for field current reference in parallel mode. In summary, the performance of the FQDC that ran in parallel mode with the assistance of LUT and powered by Genetics Algorithm sys-tem had been comparable to that of an AC drive, and thus, deemed as very suitable for applications in low-cost EV.

References

1. Westbrook, M.H.: The Electric and Hybrid Electric Vehicle, SAE (2001)
2. Husain, I.: Electric and Hybrid Electric Vehicles, Design Fundamentals. CRC Press (2003)
3. Oak Ridge National Laboratory: Advanced Brush Technology for DC Motors (2009). Available: http://peemrc.ornl.gov/projects/emdc3.jpg
4. Heinrich, Walter Rentsch, Herbert Dr.-Ing: ABB Industry: "Guide to Variable Speed Drives," Technical Guide No. 41180 D-68619 LAMPERTHEIM, Germany, 3ADW 000 059 R0201 REV B (02.01), DCS 400/DCS 500/DCS 600 (2003). Rashid, M.H.: Power Electronics, Circuits, Devices and Applications, 3rd edn., Prentice Hall (2004)
5. Arof, S., Jalil, J.A., Yaakop, N.M., Mawby, P.A., Arof, H.: Series Motor Four Quadrants Drive DC Chopper Parti: Overall. International Conference on Power Electronics (2014)
6. Arof, S., Muhd Khairulzaman, A.K., Jalil, J.A., H. Arof, Mawby, P.A.: Self tuning fuzzy logic controlling chopper operation of four quadrants drive DC chopper for low cost electric vehicle. In: International Conference on Intelligent Systems, Modeling, and Simulation, pp. 40–24. IEEE Computer Society (2015)
7. Arof, S., Zaman, M.K., Jalil, J.A., Mawby, P.A., Arof, H.: Artificial intelligence controlling chopper operation of four quadrants drive DC chopper for low cost electric vehicle. Int. J. Simul. Syst. Sci. Technol. (2015). https://doi.org/10.5013/ijssst
8. Arof, S., Jalil, J.A., Kamaruddin, N.H., Yaakop, N.M., Mawby, P.A., Arof, II.: Series motor four quadrants drive DC chopper, part 2: driving and reverse with direct current control. In: International Conference on Power Electronics, pp. 775–780. IEEE. 978-1-5090-2547-3/16 (2016). https://doi.org/10.1109/pecon2016.7951663
9. Arof, S., Hassan, H., Rosyidi, M., Mawby, P.A., Arof, H.: Implementation of series motor four quadrants drive DC chopper for DC drive electric car and LRT. J. Appl. Environ. Biol. Sci. 7(3S), 73–82 (2017)
10. Arof, S., Noor, N.M.N., Elias, F., Mawby, P.A., Arof, H.: Investigation of chopper operation of series motor four quadrants DC chopper. J. Appl. Environ. Biol. Sci. 7(3S), 49–56 (2017)
11. Arof, S., Diyanah, N.H., Mawby, P.A., Arof, H.: Study on implementation of neural network controlling four quadrants direct current chopper: part 1: using single neural network controller with binary data output. In: Advanced Engineering for Processes and Technologies, pp. 37–57 (2019)
12. Arof, S., Diyanah, N.H., Yaakop, N.M., Mawby, P.A., Arof, H.: Processor in the loop for testing series motor four quadrants drive direct current chopper for series motor driven electric car: part 1: chopper operation modes testing. In: Advanced Engineering for Processes and Technologies, pp. 59–76 (2019). https://doi.org/10.1007/978-3-030-05621-6_5
13. Negnevitsky, M.: Artificial Intelligence. A Guide to Intelligent Systems. Addison-Wesley, Essex, UK (2005)

14. Tang, X.-S., Shi, Z.-L., Li, D.-Q., Ma, L., Chen, D.: Bagging-adaboost ensemble with genetic algorithm post optimization for object detection. In: 2009 Fifth International Conference on Natural Computation, IEEE, 97 8-0-7695-3736-8/09 (2009)
15. Deniz, E., Aydogmus, O., Aydogmus, Z.: GA-based optimization and ANN-based SHEPWM generation for two level inverter. IEEE, 1-4799-7 800 (1978)

Series Motor Four Quadrants Drive DC Chopper

Part 4: Generator Mode

S. Arof, N. H. N. Diyanah, N. M. N. Noor, Md. Radzi, J. A. Jalil, P. A. Mawby and H. Arof

Abstract This paper is the part four (4) of the total 8 papers of the series motor four quadrants DC chopper that describes the generator mode operation of a four quadrant drive DC chopper that is applicable for EV traction. Alternative excitation methods proposed in this work resulted in greater build up voltage and armature current of the series motor when operating in generator mode. In order to achieve a longer traversed distance, the converter of an EV should utilize the optimum amount of power from the batteries at all time. Hence, the work has been extended to a simulation of the battery charging process and the effect due to huge voltage difference between the generated voltage and battery terminal voltage. The operation mode is modelled and simulated in MATLAB/SIMULINK and the proposed excitation methods have been verified through experimental set-up.

S. Arof (✉) · N. H. N. Diyanah · N. M. N. Noor · Md. Radzi · J. A. Jalil
Electrical Electronic Automation Section, Universiti Kuala Lumpur, Malaysian Spanish
Institute, Kulim Hi-Tech Park, 09000 Kulim, Kedah, Malaysia
e-mail: saharul@unikl.edu.my

N. H. N. Diyanah
e-mail: diyanahhisham94@gmail.com

N. M. N. Noor
e-mail: noramlee@unikl.edu.my

Md.Radzi
e-mail: mdradzi@unikl.edu.my

J. A. Jalil
e-mail: julaida@unikl.edu.my

S. Arof · P. A. Mawby
University of Warwick School of Engineering, Coventry, CV47AL, UK
e-mail: p.a.mawby@warwick.ac.uk

H. Arof
Engineering Department, Universiti Malaya, Jalan Universiti, 50603 Kuala Lumpur,
Malaysia
e-mail: ahamzah@um.edu.my

© Springer Nature Switzerland AG 2019   155
M. H. Abu Bakar et al. (eds.), *Progress in Engineering Technology*,
Advanced Structured Materials 119, https://doi.org/10.1007/978-3-030-28505-0_12

Keywords Generator · FQDC · DC drive · Series motor · Field excitation · EV · Resonator

1 Introduction

The electric vehicles (EV) and hybrid electric Vehicles (HEV) are amongst of the solutions to reduce global hydrocarbon emission. It is enthusiastically forecasted that in the near future EV will replace the internal combustion system not just because of zero emission but also due to the higher efficiency in energy conversion. The slow advancement in the development of super capacitors as alternative to batteries has always been the halt to this EV trending. On positive outlook presently, a super capacitor as big as 12,000 F has already entered the market and it is expected that much bigger storage super capacitors do exists in research laboratories in certain R&D rigorous regions around the world [1]. The only drawback is that the super capacitor operates at lower voltage (as low as 2.7 V), nonetheless, this is will not be an ongoing disadvantage for any DC drive as DC drives nowadays are trending towards low operating voltage motors [2].

The Oak Research National Laboratory (ORNL), a US based research center, in 2003 had succeeded to develop a 55 kW, high efficiency brushed motor (92%) that operates at 13 V. It is expected that currently ORNL is producing lower than 5 V operating voltage to suit the super capacitor to match electric car applications. Such motors possess high power output, higher efficiency, smaller size, less weight, long lasting carbon brush and commutator, low operating voltage and use a modular structure. Nowadays, the lifespan of the carbon brush and commutator of a DC motor is longer than the rotor bearing of an AC induction motor. The brush can last until 30,000 km while the commutator can endure 250,000 km before it flushes over. The reliability of the soft-commutated DC motor is now around 90–94%, while the spark reduction system can further extend the brush life [2–6].

2 Four Quadrants Drive DC Chopper Operation Modes

The DC choppers were introduced in the early 1960s using force-commutated thyristor. DC choppers are mainly used to drive DC motors while offering the capability of bidirectional power flow for both motoring and regenerative braking. For EV application using series motor, the common half-bridge DC chopper shown in Fig. 1 offers no capability to perform reverse, field weakening and resistive braking [7–9]. The generator and regenerative braking mode depend on the residual magnetic field. Alternatively, a four quadrants DC chopper (FQDC) proposed in [4] offers seven modes of operation; i.e. forward and reverse driving, field weakening, parallel driving, generating, regenerative braking, and resistive braking capabilities which is depicted in Fig. 2.

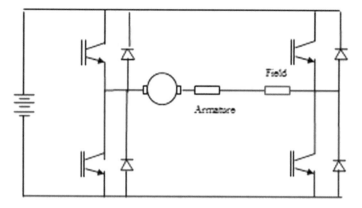

Fig. 1 H-bridge chopper

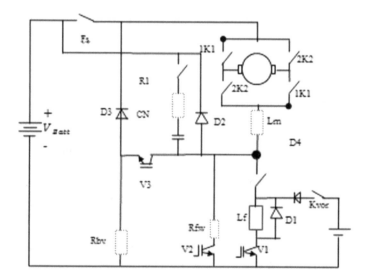

Fig. 2 Four quadrants drive DC chopper for series motor

This modes of operation can be controlled by AI controllers such as expert system, fuzzy logic, self tuning fuzzy [10–16], neural network as well as ANFIS. But, this paper focused only on the circuit operation of generator mode. The chopper has the main IGBT $V1$, field weakening IGBT $V2$ and bridging IGBT $V3$. L_M is the motor inductance connected in series with the armature windings which is used to reduce the armature ripple current, while R_{BV} is the brake series resistor. Equations (1)–(5) describes the voltage and current for the chopper and is applicable to all chopper operation modes. B_{emf} is the back emf of the motor, V_a and V_f are the armature and field voltages respectively, K_b is the back emf constant, K_t is the motor torque constant, I_f is the field current, R_a and R_f are the motor coil resistances, I_a is the armature current, ω is the angular speed, while T_d is the motor torque. Hence, this paper is aimed to discuss on different methods of excitation for the generator mode of the FQDC proposed in [4] to achieve better vehicle speed control as the vehicle is moving downhill.

$$I_a = \frac{V_{batt} - I_a\left(R_a + R_f\right) - B_{emf}}{L_a + L_f} \tag{1}$$

$$B_{emf} = K_v I_f \omega \tag{2}$$

$$T_d = K_t I_f I_a = \mathrm{J}\frac{d\omega}{dt} + B_w + T_L \tag{3}$$

2.1 Methods of Excitation for Generator Mode

The EV requires power to charge the batteries. The excess kinetic energy available while the vehicle is moving downhill can be used to generate electrical power to charge the batteries. The generated power can be collected at the armature winding. Therefore, the field winding has to be charged with electrical energy to establish the magnetic field. When the contactor K_{vor} is closed, IGBT $V1$ is fired up to allow current to flow into the field winding. This action causes the armature winding to start generating a build-up voltage and current [17, 18]. As the field winding is excited by means of the auxiliary source, voltage is generated at the armature. An increase in field current will boost the generated voltage at constant speed [19, 20]. During generator mode, both the armature and field currents can be represented by Eqs. (4) and (5). This method of excitation is commonly applied in any series DC motor as described in Fig. 3.

When contactor $2K2$ is energized, current I_a changes its direction thus becoming negative as described by Eq. (6). The armature current is negative because the current direction is opposite while in driving, field weakening and parallel operation mode. Gradually the armature voltage A_v starts to build up. This voltage is used to charge the batteries.

$$\frac{d}{dt}[I_f] = \left[-\left(\frac{R_f}{L_f}\right)\right][I_f]\left[\frac{1}{(L_f)}\right][V_{ext}K_{v1}] \tag{4}$$

$$\frac{d}{dt}[I_a] = \left[\frac{-(R_{bh} + R_{batt} + R_a)}{L_a}\right][I_a]\left[\frac{1}{L_a}\right][B_{emf}] \tag{5}$$

$$I_{batt} = -I_a \tag{6}$$

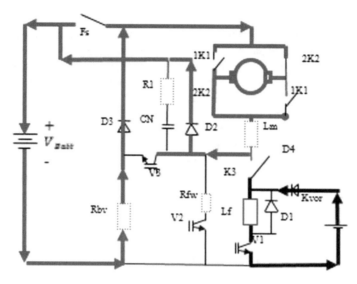

Fig. 3 Field excitation by auxiliary supply

The FQDC also offers an alternative excitation method for the generator mode without constantly utilizing auxiliary power. This method is very useful if the EV is moving downhill. As contactor *K3* and contactor *Kvor* are switched on while IGBT *V1* is fired at this moment, the field winding is pre-excited via the auxiliary battery as shown in Fig. 4.

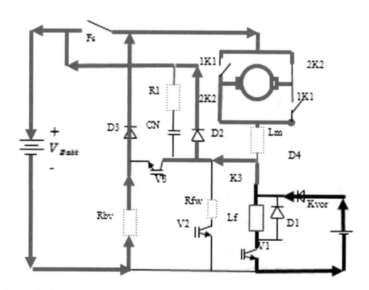

Fig. 4 Pre-excitation by auxiliary supply

Consequently, EV speed increases as it moves downhill, provided that no mechanical braking action is applied. When the voltage generated at the armature is higher than the auxiliary battery voltage, it causes reverse biased to diode *V4*. Thus, current from the auxiliary battery can no longer flow to the field winding. The current that flows to the field winding is mainly provided by the generated voltage, therefore contactor *Kvor* can be switched off.

As an alternative to the two excitation methods discussed earlier, the field winding can also be pre-excited by firing the resonator circuit as shown in Fig. 5. The voltage and current for the resonator circuit are described by Eqs. (7)–(11). When the resonator capacitance is fully charged, current will stop flowing to the resonator circuit. As the rotor keeps rotating while EV is moving downhill, voltage starts to build up in the armature winding. When the generated voltage is greater than the battery voltage, armature current will flow to the battery via diode D5 and start charging the battery as shown in Fig. 6. This excitation method offers a higher armature current to charge the battery.

The armature current is represented by Eq. (12).

$$V_{dc} = L_f \frac{di}{dt} + I(R_c + R_f) + V_c \tag{7}$$

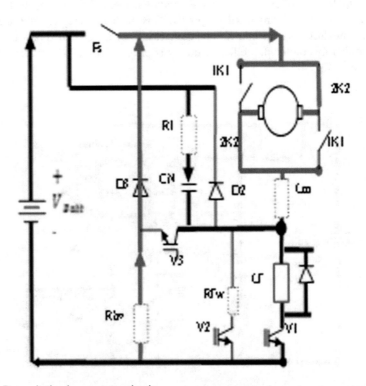

Fig. 5 Pre-excitation by resonator circuit

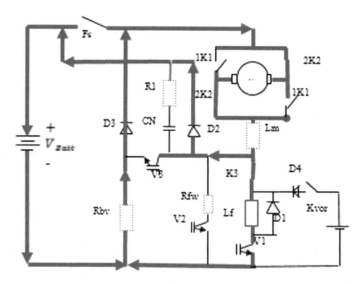

Fig. 6 Path of current upon removal of pre-exiter

$$V_c = \frac{1}{C} \int i dt \qquad (8)$$

$$\frac{dV_c}{dt} = \frac{I}{C} \qquad (9)$$

$$I = C \frac{dV_c}{dt} \qquad (10)$$

$$L_f C \frac{d^2 V_c}{dt} + (R_f + R_c) C \frac{RCV_c}{dt} + V_c = V_{dc} \qquad (11)$$

$$I_a = I_{batt} + I_{field} \qquad (12)$$

In order to charge the battery, the generated voltage must be slightly higher than the battery voltage. Consequently, the generated current is always proportional to the difference between battery terminal voltage and chopper generated voltage. A huge difference between the two voltages may develop a higher charging current which leads to a rise in battery temperature. Consistently increasing the battery temperature may reduce its life time, eventually causes damage to the battery. The battery current is equivalent to the generated armature current but in negative direction. Since the armature current has a higher negative magnitude, a higher generator torque is developed thus produces jerk to the EV. Equations (13) and (14) represent the battery current and generator torque, respectively.

$$I_{batt} = -I_a \qquad (13)$$

$$Tq_{gen} = Kt(-I_a)I_f \qquad (14)$$

2.2 Experimental Setup

In order to test the FQDC in generator mode, an experimental set-up as shown in Figs. 7 and 8 were prepared. A 650 W series wound DC motor shaft is coupled with another motor with inertia load (flywheel) to represent a downhill movement of an EV.

The expected charging voltage is set according to the battery state of charge. The value of voltage needed to be charged is sent to the PIC 18F45K22 microcontroller. The appropriate charging voltage is selected from a look-up table based on the present state of charge value. The selected voltage is then fed to a PID controller. The PID algorithm will compute the data and feed the output to the PIC. A set of pulse width modulation (PWM) signal is then fed to the IGBTV1 of the FQDC. Experimental results were captured using a data acquisition software. The experimental values of motor speed, armature current, field current and armature voltage were transferred to the FQDC model in MATLAB/SIMULINK to validate the simulation model. A block diagram of the FQDC operating in generator mode is represented in Fig. 9.

Fig. 7 Four quadrants drive DC chopper for series motor

Fig. 8 The experimental set-up for the chopper

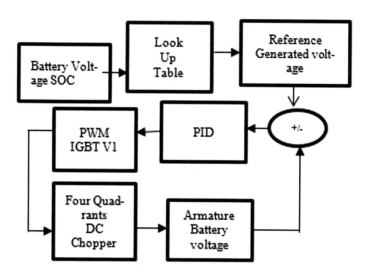

Fig. 9 Block diagram for generator mode

3 Results and Discussion

The effects on voltage, current and speed for the three excitation methods of the FQDC operating in generator mode are presented in Figs. 10, 11 and 12.

When the field winding is fully excited by an auxiliary supply as shown in Fig. 10, voltage is generated at the armature winding. At this moment, the speed is observed to be constant since voltage at the armature terminal is not connected to any load. As the controller triggers a generator signal, the FQDC starts to operate on generator mode. Consequently, more voltage is generated within the armature winding. As load is applied, a negative torque is developed and causes the speed to drop. Hence, a reduction in speed leads to a drop in the generated voltage. A decrease in the armature current is also observed due to the effect of the negative torque.

Figure 11 shows the effect on voltage, current and speed as the field winding is pre-excited by an auxiliary supply. In this method, the armature voltage is observed to be greater than the auxiliary battery voltage as load is applied at its terminal. Thus allowing more current flow to the field winding circuit and causes an increase in the field winding current. The motor speed drops as a result of the negative torque. It is also observed that the speed drop is higher as compared to the one in the common excitation method. The sum of both field and battery currents resulted an increase in the armature current.

In an alternative excitation technique, the field winding is pre-excited by a resonator circuit as shown in Fig. 12 so that the armature coil could induce the required generated voltage. The generated voltage is fed to field winding which causes an increase in armature current. As a result, a change in the motor speed is observed. When load is applied, the motor speed drops due to the negative torque

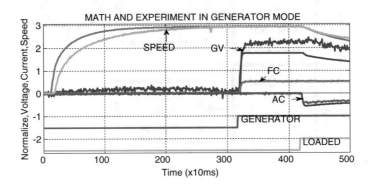

Fig. 10 Voltage, current and speed of field excitation by auxiliary supply

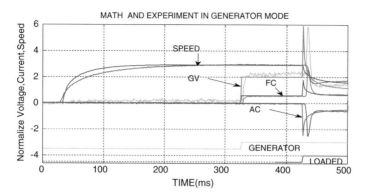

Fig. 11 Voltage, current and speed of pre-excitation by auxiliary supply

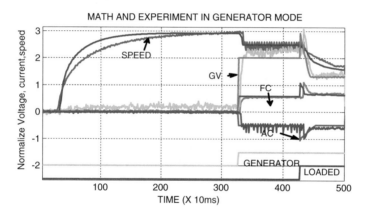

Fig. 12 Voltage, current and speed of pre-excitation by resonator circuit

developed at the rotor. It is also observed that the speed drop is higher than the one in the common excitation technique due to an increase in the armature current. The armature current is increased due to the sum of both field and battery currents that flow in the armature circuit.

As speed decreases due to the negative torque which counters the rotation direction of the rotor shaft, the generated voltage is gradually increased to the required charging voltage as simulated in Fig. 13. The required charging voltage is regulated in order to maintain the charging current. At 100% state of charge (SOC), the battery voltage remains at its saturation value which indicates that the battery is fully charged.

Fig. 13 Simulation of speed, voltage and battery SOC

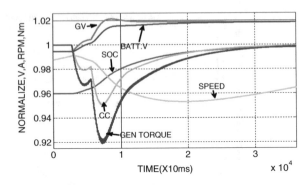

4 Conclusions

The generator mode of the four quadrants drive DC chopper has been modeled and simulated. The experimental results revealed that it is capable of enhancing dc series motor and has high potential to be utilized in EV. The generator mode can be successfully achieved by utilizing two excitation methods proposed in this work as an alternative to the common excitation method applied for series motor. In both alternative methods, the field winding is momentarily pre-excited by an auxiliary supply or a resonator circuit. Hence, the power consumption from the auxiliary source is minimized. The alternative methods also offer better speed control as the EV moves downhill. An electrical braking is activated as the rotor developed higher negative torque due to an appropriate selection of the chopper operation mode. In summary, the performance of an optimized DC drive system with the generator mode feature, do offer favorable advantages to that of an AC drive and thus is very suitable for applications in this cost effective EV solution.

References

1. Alcicek, G., Gualous, H., Venet, P., Gallay, R., Miraoui, A.: Experimental study of temperature effect on ultracapacitor ageing. In: Power Electronics and Application, European Conference (2007)
2. Oak Ridge National Laboratory: Advanced Brush Technology for DC Motors (2009). Available: http://peemrc.ornl.gov/projects/emdc3.jpg
3. Oak Ridge National Laboratory: Soft-Commutated Direct Current (DC) Motor (2009). Available: www.ornl.gov/etd/peemrc
4. Arof, S., Yaakob, N.M., Jalil, J.A., Mawby, P.A., Arof, H.: Series motor four quadrants drive DC chopper, Part 1: overall. In: International Conference on Power Electronics (2014)
5. Heinrich, Walter Rentsch, Herbert Dr.-Ing: ABB Industry: Guide to Variable Speed Drives. Technical Guide No. 41180 D-68619 LAMPERTHEIM, Germany, 3ADW 000 059 R0201 REV B (02.01), DCS 400/DCS 500/DCS 600: ABB (2003)
6. Westbrook, M.H.: The Electric and Hybrid Electric Vehicle. SAE (2001)
7. Husain, I.: Electric and Hybrid Electric Vehicles, Design Fundamentals, CRC Press (2003)

8. Rashid, M.H.: Power Electronics, Circuits, Devices and Applications, 3rd edn. Prentice Hall (2004)
9. Bansal, R.C.: Birla Institute of Technology and Science, Pilani, India. In: Electric Vehicle. Handbook of Automotive Power Electronics and Motor Drives, Taylor & Francise Group CRC Press (2005)
10. Arof, S., Muhd Khairulzaman, A.K., Jalil, J.A., Arof, H., Mawby, P.A.: Self tuning fuzzy logic controlling chopper operation of four quadrants drive DC chopper for low cost electric vehicle. In: 6th International Conference on Intelligent Systems, Modeling and Simulation, pp. 40–24 (2015)
11. Arof, S., Muhd Khairulzaman, A.K., Jalil, J.A., Arof, H., Mawby, P.A.: Artificial intelligence controlling chopper operation of four quadrants drive DC chopper for low cost electric vehicle. Int. J. Simul. Sci. Technol. (2015). https://doi.org/10.5013/ijssst.a.16.04.03,2015.ijsst
12. Arof, S., Jalil, J.A., Kamaruddin, N.H., Yaakop, N.M., Mawby, P.A., Arof, H.: Series motor four quadrants drive DC chopper, part 2: driving and reverse with direct current control. In: International Conference on Power Electronics, pp. 775–780, 978-1-5090-2547-3/16 (2016). https://doi.org/10.1109/pecon2016.7951663
13. Arof, S., Hassan, H., Rosyidi, M., Mawby, P.A., Arof, H.: Implementation of series motor four quadrants drive DC chopper for DC drive electric car and LRT. J. Appl. Environ. Biol. Sci. 7(3S), 73–82 (2017)
14. Arof, S., Noor, N.M.N., Elias, F., Mawby, P.A., Arof, H.: Investigation of chopper operation of series motor four quadrants DC chopper. J. Appl. Environ. Biol. Sci. 7(3S), 49–56 (2017)
15. Arof, S. Diyanah, N.H.N., Mawby, P.A., Arof, H.: Study on implementation of neural network controlling four quadrants direct current chopper: part 1: using single neural network controller with binary data output. In: Advanced Engineering for Processes and Technologies, pp. 37–57 (2019)
16. Arof, S., Diyanah, N.H.N., Yaakop, M.N., Mawby, P.A., Arof, H.: Processor in the loop for testing series motor four quadrants drive direct current chopper for series motor driven electric car: part1: chopper operation modes testing. In: Advanced Engineering for Processes and Technologies, pp. 59–76 (2019). https://doi.org/10.1007/978-3-030-05621-6_5
17. Ching, T.W.: Soft-switching converters for EV propulsion. J. Asian Electr. Veh. 5(2), 1019–1026 (2007)
18. Senthil Kumar, N., Sasasivam, V., Muruganandam, M.: A low cost four-quadrant chopper—fed embedded DC drive using fuzzy controller. J. Electr. Power Compon. Syst. 35, 907–920 (2007)
19. Joshi, D., Bansal, R.C.: Performance evaluation of multiquadrant DC drive using fuzzy-genetic approach. J. Electr. Syst. (JES) 5(4), 1–9 (2009)
20. Tipsumanporn, V., Thepsathorn, P., Piyarat, W., Numsomran, A., Bunjungjit, S.: 4-quadrant DC motor drive control by BRM technique. In: Warwick University—IEEE International Conferences on Automation, Quality and Testing, Robotics, pp. 265–270 (2007)

Relationship Between Electrical Conductivity and Total Dissolved Solids as Water Quality Parameter in Teluk Lipat by Using Regression Analysis

Nor Haniza Bakhtiar Jemily, Fathinul Najib Ahmad Sa'ad, Abd Rahman Mat Amin, Muhammad Firdaus Othman and M. Z. Mohd Yusoff

Abstract This study investigated the relationship between electrical conductivity and total dissolved solids (TDS) as water quality parameters at Teluk Lipat coastal area. Teluk Lipat is located at the East coast of Malaysia peninsula that is directly exposed to the South China Sea. 13 water samples from this area were collected to determine of electrical conductivity (EC) and total dissolved solids (TDS). From the regression analysis between electrical conductivity (EC) and total dissolved solids (TDS), it shows a very strong relationship between these two parameters with R^2 value of 0.9306 but it was still not a perfect straight line. This analysis can be used to give an over-view of water quality.

Keywords Electrical conductivity · Total dissolved solid (TDS) · Coastal area · Regression

N. H. Bakhtiar Jemily
Mechanical Section, Universiti Kuala Lumpur Malaysian Spanish Institute,
Kulim Hi-Tech Park, 09000 Kulim, Kedah, Malaysia
e-mail: norhaniza@unikl.edu.my

F. N. Ahmad Sa'ad (✉) · A. R. Mat Amin
Faculty of Applied Science, Universiti Teknology Mara,
23200 Bukit Besi, Terengganu, Malaysia
e-mail: najib4496@yahoo.com

A. R. Mat Amin
e-mail: abdra401@tganu.uitm.edu.my

M. F. Othman · M. Z. Mohd Yusoff
Department of Applied Science, Universiti Teknologi Mara,
13500 Permatang Pauh, Pulau Pinang, Malaysia
e-mail: firdaus327@ppinang.uitm.edu.my

M. Z. Mohd Yusoff
e-mail: zaki7231@ppinang.uitm.edu.my

© Springer Nature Switzerland AG 2019
M. H. Abu Bakar et al. (eds.), *Progress in Engineering Technology*,
Advanced Structured Materials 119, https://doi.org/10.1007/978-3-030-28505-0_13

1   Introduction

Electrical conductivity (EC) and total dissolved solids (TDS) are normally used as water quality parameters, especially in the seawater. These two parameters are indicators of salinity level which make these parameters very useful in studying seawater invasion [1–5]. Matter present in the dissolved form consists of inorganic salts and organic matter which is represented in the form of total dissolved solids [6] and EC is the measure of liquid capacity to conduct an electric charge [7–9]. The relationship between EC and TDS is a function of the type and nature of the dissolved cations and anions in the water [10]. Therefore, researchers have done various investigations to find out the precise mathematical correlation between these two parameters, so TDS concentration can be simply calculated from the EC value. The EC-TDS relationship is affected by the ionic composition of the water and the concentration of dissolved species. This study will investigate the relationship between electrical conductivity and total dissolved solids as water quality parameters at Teluk Lipat.

2   Study Area

Teluk Lipat is located in Dungun, Terengganu. This beach stretches about 13 km to Kuala Dungun from Bukit Bauk that is located at the South. Along this line are located a University, a school, chalets and residential houses. The road along this coastal is really close to the beach. Figure 1 shows the study area.

3   Methodology

A field trip was conducted to collect the water quality parameter along the study area on 28 March 2018. From this trip, 13 samples for saline water were carried out. The distance between each station was 1 km. Then the samples have been analyzed in the laboratory. Table 1 shows the values for electrical conductivity and total dissolved solids from 13 stations.

A linear regression analysis was used to developed predictive models for electrical conductivity and total dissolved solids. The regression analysis was performed using the data of electrical conductivity and total dissolved solids. For the regression model development, the total samples (n = 13) were used for regression modelling.

Fig. 1 Study area

Table 1 Values for electrical conductivity and total dissolved solids

Station	Total dissolved solid (mg/L)	Electrical conductivity (µS/cm)
St1	24,167	40,385
St2	25,018.5	41,630
St3	24,687	40,954
St4	24,895	41,437
St5	24,908	41,485
St6	24,908	41,435
St7	24,836.5	41,272
St8	25,005.5	41,634
St9	24,765	41,287
St10	24,979.5	41,956
St11	25,272	42,105
St12	25,317.5	42,013
St13	25,265.5	42,093

Fig. 2 The EC and TDS
correlation in a linear
regression analysis

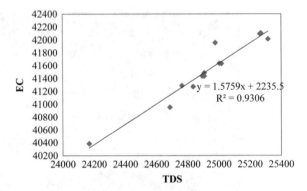

4 Result and Discussion

Figure 2 represents the EC and TDS correlation in a linear regression analysis. From this result, the value of coefficient (R^2) is 0.9306. By conducting a linear equation, the mathematical formula between EC and TDS in seawater will fit the following equation.

$$EC = 1.5759\,(TDS) + 2235.5 \quad (1)$$

From this regression analysis between electrical conductivity and total dissolved solids, it shows a very strong relationship between these two parameters but it was still not a perfect straight line.

5 Conclusion

EC and TDS are water quality parameters which indicate the level of salinity. The measurement of EC value is far easier than the one of TDS. Meanwhile obtaining TDS concentration is principal because it can explain the water quality in a more complex manner than only from the EC value. From the result, the relationship between EC and TDS is not always linear in this study area. This situation highly depends on the water salinity and material contents. The acquisition of EC from TDS conversion can be conducted in explaining general condition of water quality.

References

1. Rusydi, A.F.: Correlation between conductivity and total dissolved solid in various type of water: a review. In: IOP Conference Series: Earth and Environmental Science, vol. 118(1), p. 012019. IOP Publishing, Feb 2018
2. El Moujabber, M., Samra, B.B., Darwish, T., Atallah, T.: Comparison of different indicators for groundwater contamination by seawater intrusion on the Lebanese coast. Water Resour. Manage **20**(2), 161–180 (2006)
3. Stigter, T.Y., Ribeiro, L., Dill, A.C.: Application of a groundwater quality index as an assessment and communication tool in agro-environmental policies–two Portuguese case studies. J. Hydrol. **327**(3–4), 578–591 (2006)
4. Nonner, J.C.: Introduction to Hydrogeology. CRC Press, Taylor and Francis Group, London (2015)
5. Han, D., Kohfahl, C., Song, X., Xiao, G., Yang, J.: Geochemical and isotopic evidence for palaeo-seawater intrusion into the south coast aquifer of Laizhou Bay, China. Appl. Geochem. **26**(5), 863–883 (2011)
6. Thirumalini, S., Joseph, K.: Correlation between electrical conductivity and total dissolved solids in natural waters. Malays. J. Sci. **28**(1), 55–61 (2009)
7. Marandi, A., Polikarpus, M., Jõeleht, A.: A new approach for describing the relationship between electrical conductivity and major anion concentration in natural waters. Appl. Geochem. **38**, 103–109 (2013)
8. Daniels, W.L., Zipper, C.E., Orndorff, Z.W., Skousen, J., Barton, C.D., McDonald, L.M., Beck, M.A.: Predicting total dissolved solids release from central Appalachian coal mine spoils. Environ. Pollut. **216**, 371–379 (2016)
9. Logeshkumaran, A., Magesh, N.S., Godson, P.S., Chandrasekar, N.: Hydro-geochemistry and application of water quality index (WQI) for groundwater quality assessment, Anna Nagar, part of Chennai City, Tamil Nadu, India. Appl. Water Sci. **5**(4), 335–343 (2015)
10. Sawyer, C.N., McCarty, P.L., Parkin, G.F.: Chemistry for Environmental Engineers. Mc Graw-Hill Book Company, New York (1978)

A Study of the Region Covariance Descriptor: Impact of Feature Selection and Precise Localization of Target

Mohd Fauzi Abu Hassan, Azurahisham Sah Pri, Zakiah Ahmad and Tengku Mohd Azahar Tuan Dir

Abstract In visual tracking, selecting the right image descriptors is critical. The popular version of descriptor is known as the covariance descriptor; however, no further studies is yet developed regarding the different methodologies for its construction. This study analyzes the contribution of diverse features of an image to the descriptor and their contribution to the detection of arbitrary targets in sequences of images, in our case: *Boy, David3, Bolt* and *Walking2* in an image sequence. The methodology to determine the performance of the covariance matrix is defined from different sets of characteristics, and a specific combination of features is needed to develop a correlation between them. Finally, when an analysis is performed with the best set of features, F4 the *target* detection problem reached a performance of average, 0.77, From this experiment, it is believed that we have constructed a greater solution in choosing best features for this descriptor, allowing to move forward to the next issues such as using it on others datasets.

Keywords Object detection · Covariance descriptor · Feature selection

M. F. Abu Hassan (✉) · A. Sah Pri · Z. Ahmad
Universiti Kuala Lumpur Malaysian Spanish Institute, Kulim Hi-Tech Park,
09000 Kulim, Kedah, Malaysia
e-mail: mohdfauzi@unikl.edu.my

A. Sah Pri
e-mail: azurahisham@unikl.edu.my

Z. Ahmad
e-mail: zakiah@unikl.edu.my

T. M. A. Tuan Dir
Universiti Kuala Lumpur Malaysian France Institute, Section 14, Jalan Teras Jernang,
Seksyen 14, 43650 Bandar Baru Bangi, Selangor, Malaysia
e-mail: tgazahar@unikl.edu.my

© Springer Nature Switzerland AG 2019
M. H. Abu Bakar et al. (eds.), *Progress in Engineering Technology*,
Advanced Structured Materials 119, https://doi.org/10.1007/978-3-030-28505-0_14

1 Introduction

Feature selection is one of the most important steps for detection and classification problems. Good features should be discriminative, robust, easy to compute and efficient algorithms are needed for a variety of tasks such as recognition and tracking. [1]. A feature is a piece of information that describes an image or a part of it e.g. color, edges etc. Features are useful to match two images. Properties of good features are repeatability, saliency and compactness.

Color stays constant under geometric transformations. Color is a local feature which it is defined for each pixel and robust to partial occlusion. The ideas are that it can use object colors directly for recognition, or better—use statistics of object colors.

In this paper, we have defined a methodology which determinates the performances of different covariance matrices built from distinct sets of features. With this we are able to define which ones are the best for detection of any objects. First, we obtain a set of images, where we select a specific target that we want to detect. Next, we find, in an image search, the region with the smallest distance to the target region initially selected. We obtain the performance for each set of features used in the creation of the covariance descriptor. Finally, when an analysis with the best set of features is performed we get for the object the localization problem.

In the object detection field, there are many different approaches, one of them is based on features. In this approach, we identify two main tasks. The first one is to extract the features which should give as much information as they can from the object, region or image. The second task is the detection of the object or region through a good classification of the features previously extracted.

The remainder of this paper is organized as follows. In Sect. 2, we review related work based on a covariance descriptor framework. In Sect. 3, we describe our approach. In Sect. 4, we show experimental results. Finally, we present our conclusions.

2 Related Work

Most applications that employ the covariance descriptor compute the descriptor using a fixed set of features, which is often determined a priori. For instance, in [1], each pixel is converted to a nine-dimensional feature vector for object detection: This feature set remains unchanged in [1] for all kinds of objects, without considering the characteristics of each object.

Actually, color can be interpreted and modeled in different ways. With the availability of a variety of color spaces, e.g. RGB, HSV, YCrCb, YUV, CIE Lab, CIE Luv, etc., the inevitable question is how to select proper color models that can produce good performance for detecting a particular object. Likewise, the gradient features, which encode the shape information of the region context, can also have a variety of choices. Indeed, they can be computed using different combinations of orders, and further with their corresponding magnitudes and orientations. Consequently, how to choose the feature set to be fused in the covariance descriptor for detection is of great importance.

A number of works [2–4] have empirically studied the performances of the covariance descriptor using different feature sets. The reported results showed that significantly different performances were achieved when using different features. This further shows the importance of feature selection or extraction for the covariance descriptor.

Alahi et al. [2, 5] compared different feature sets for detection and tracking objects across non-calibrated camera networks and claimed that increasing the number of features may increase the performance of covariance descriptor. In addition, Alahi et al. suggested that shape information is crucial for inter-category object detection. For instance, gradient features perform well in pedestrian detection applications because the shape of a human is a relevant cue to distinguish it from other objects, whereas color features perform best in intra-category classification cases such as object re-identification or recognition. In [3], Cortez-Cargill et al. constructed covariance descriptors with nine sets of features based on various color spaces. They obtained a best feature set which embeds many color channels from a variety of color spaces and reaches a performance of 99% for face detection. However, the feature vector they got turned out to be a 20-dimensional one and thus makes the construction and similarity measure of the covariance matrices rather time-consuming.

In brief, two points can be drawn. First, different feature combinations generally produce different detection performances. Second, previous works generally reported better results using more features. Subsequently, two questions naturally arise. First, how to select a proper feature set for detecting a specific object to ensure good performance in terms of detection accuracy? Second, is it always true that fusing more features produces better detection performance? If yes, are there alternatives that use compact feature set while ensuring good performance? If not, what is the condition when more features do not yield better performance? We will try to answer these questions in the next section by analyzing the generalization ability of the region covariance descriptor from a machine learning perspective.

3 Our Approach

3.1 Region Covariance Descriptor

For the appearance model, we use the region covariance descriptor [1, 3] as the observation model. The region covariance descriptor can describe the object more accurately by combining different types of spatial and temporal features naturally.

Since the dimension of the covariance matrix depends only on the number of the features we used, the problem of high calculation cost in high dimension is avoided.

In this work, we select the feature vector as

$$F(i) = [f1, f2, f3 \ldots fn], \tag{1}$$

where $\{fi\}_i^m$ are d-dimensional feature column vectors associated with the pixel index i, and m is the number of total pixels in the image. Feature vectors are generated for each pixel.

Then the covariance matrix is calculated by

$$Cr = \frac{1}{MN} \sum_{k=1}^{MN} (fk - \mu r)^t (fk - \mu r)^T \tag{2}$$

where r indicates the rectangle region with the scale: MN, and μr is the mean of all vectors in the region r. In this way, the region r is represented as a MN covariance matrix. Since covariance matrices lie on the Riemannian space rather than the Euclidean space, the Bhattacharyya distance, which is often used for histogram distance metric, cannot be used to measure the distance of covariance matrices. Therefore, the dissimilarity between two covariance matrices is proposed in [1] and calculated by

$$\rho(Ci, Cj) = \sqrt{\sum_{k=1}^{d} ln^2 \lambda k(Ci, Cj)}, \tag{3}$$

where Ci and Cj are covariance matrices of the target model and the candidate particle, respectively; λk is the generalized eigenvalue between Ci and Cj. Lastly, find the target region with minimum distance [1, 3] as the matching region.

$$\rho min = argmin\, \rho(Ci, Cj), \tag{4}$$

where ρ is the covariance matrix distance.

The right features are exactly those that make the tracker work best. Color descriptors have been successful in many computer vision applications. Here, we investigate them for the visual tracking problem. Figure 1 shows flow chart of the covariance tracker used in this experiment.

Finally, we used six different potential features to build the covariance descriptor (Table 1). Table 2 shows the description of each features.

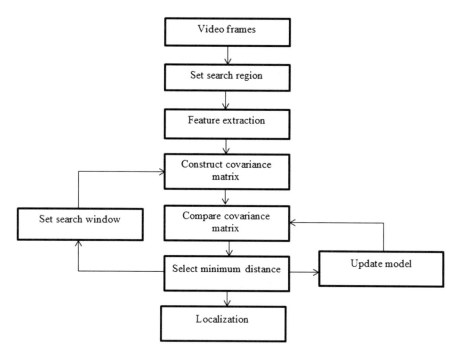

Fig. 1 Flow chart of covariance tracker

Table 1 Potential features

F1	R	G	B						
F2	R	G	B	E					
F3	R	G	B	V	H				
F4	R	G	B	V	H	E			
F5	R	G	B	V	H	E	L		
F6	R	G	B	V	H	E	Q	S	U

Table 2 Description of potential features

	Description
R	R values of the pixel
G	G values of the pixel
B	B values of the pixel
V	image intensity first derivative in *vertical*
H	image intensity first derivative in *horizontal*
E	absolute value V and H
L	second derivative's magnitude
Q	image intensity second derivative in *vertical*
S	image intensity second derivative in *horizontal*
U	image intensity second derivative in mixed *derivative*

4 Experiments and Comparison

In this section, we present quantitative tracking results. Our experiment is conducted on the CVPR2013 tracking benchmark [6], which is specially designed for evaluation of tracking performance. The dataset includes 36 RGB sequences and 15 grayscale sequences.

4.1 Evaluation Methodology

We use the success rate and precision for quantitative evaluations, which are common evaluation criteria. *Precision*: Percentage of frames where the predicted location is within 20 pixels of the ground truth as a tracking evaluation method is widely used. It is defined as the Euclidean distance between the target center and the ground truth. The precision score shows the percentage of frames whose it is within some given threshold. The default threshold is equal to 20 pixels.

4.2 Feature Selection

In the following feature description, we refer to some published articles [1–3]. Tables 1 and 2 show the potential feature and their description respectively. We first perform an experiment to select optimal features for the covariance descriptor. The performance of the features descriptors mentioned before is evaluated on the task of object tracking. Table 3 shows the results on selected videos of the CVPR2013 tracking benchmark. The results are reported in precision at a threshold of 20 pixels of the ground truth.

Table 3 Tracker precision at a threshold of 20 (percentage of frames where the predicted location is within 20 pixels of the ground truth)

Video sequences	F1	F2	F3	F4	F5	F6
Boy	0.78	0.86	0.82	0.98	0.98	0.91
David3	NaN	NaN	0.82	0.83	0.81	0.74
Bolt	0.02	0.60	0.69	0.75	0.02	0.71
Walking2	0.43	0.41	0.48	0.51	0.58	0.52
Average	**0.41**	**0.62**	**0.70**	*0.77*	**0.6**	*0.72*

This threshold was used by Babenko et al. [7]. In each row best result is presented by blue color and second best by red color
*NaN denotes the tracker loses the target for several frames during tracking

4.3 Quantitative Analysis

All six (6) features were used in *Boy, David3, Bolt* and *Walking2* databases. At the end, all features selected were generate data trajectory rescpectively. All six can be seen in Fig. 2. These are *F1, F2, F3, F4, F5* and *F6*.

Boy. Tracking Object Location: Precision at a fixed threshold of 20. Covariance tracker used *F4* and *F5* indicates best performance.

David3. At the end, all features selected were produced data trajectory respectively except *F1* and *F2*. No data trajectories were generated because of the tracker loses the target during tracking. Tracking Object Location: Precision at a fixed threshold of 20. Covariance tracker used *F4* indicates best performance.

Bolt. Tracking Object Location: Precision at a fixed threshold of 20. Covariance tracker used *F4* indicates best performance.

Walking2. Tracking Object Location: Precision at a fixed threshold of 20. Covariance tracker used *F5* indicates best performance.

The summary of tracker performance on many sequences is shown in Table 3. The average tracker precision shows that *F4* indicates best performance.

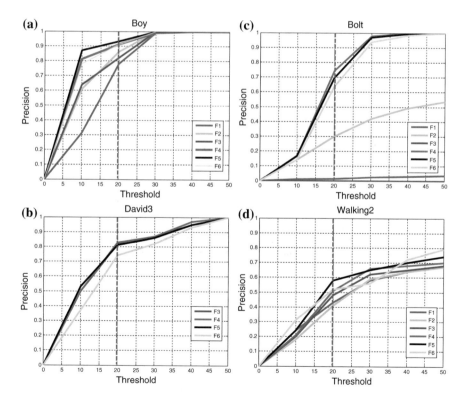

Fig. 2 Precisions plots for 4 sequences (percentage of frames where the predicted location is within the threshold of the ground truth). **a** Boy, **b** David3, **c** Bolt, **d** Walking2

5 Conclusions

From the case studies it could be concluded that the covariance descriptor has various aspects to be considered. We discussed six different features that are commonly used for their significance. We also explored the precision of the feature selection through experiments in which we integrated several feature extractions. Our findings that *F4* is best for all datasets, and that the choice of feature can have impact on performance, which lead a significant increase in performance for object tracking. In our future work, we can select *F4* as our feature on covariance tracker. The goal of our next approach is to minimize drifting and improve the tracking performance of the covariance tracker. We believe that our method can also obtain excellent results for other computer vision tasks including object tracking and so on. We hope this work can inspire more researchers continuously to study this technique and expand it on designing robust models.

References

1. Tuzel, O., Porikli, F., Meer, P.: Region covariance: a fast descriptor for detection and classification. In: Proceedings of 9th European Conference on Computer Vision, vol. 2, pp. 589–600 (2006)
2. Alahi, A., Vandergheynst, P., Bierlaire, M., Kunt, M.: Cascade of descriptors to detect and track objects across any network of cameras. Comput. Vis. Image Underst. **114**(6), 624–640 (2010)
3. Cortez-Cargill, P., Undurraga-Rius, C., Mery-Quiroz, D., Soto, A.: Performance evaluation of the covariance descriptor for target detection. In: International Conference of the Chilean Computer Science Society, pp. 133–141 (2009)
4. Metternich, M.J., Worring, M., Smeulders, A.W.M.: Color based tracing in real-life surveillance data. In: Transactions on Data Hiding and Multimedia Security. Springer, Berlin (2010)
5. Alahi, A., Marimon, D., Bierlaire, M., Kunt, M.: Object detection and matching with mobile cameras collaboration with fixed cameras. In: Proceedings of 10th European Conference on Computer Vision, pp. 1523–1550 (2008)
6. Wu, Y., Lim, J., Yang, M.: Online object tracking: a benchmark. In: CVPR (2013)
7. Babenko, B., Yang, M., Belongie, S.: Robust object tracking with online multiple instance learning. IEEE Trans. Pattern Anal. Mach. Intell. **33**(8), 1619–1632 (2011)

Analysis of a Micro Francis Turbine Blade

K. Shahril, A. Tajul, M. S. M. Sidik, K. A. Shamsuddin
and A. R. Ab-Kadir

Abstract The main purpose of the turbine is to extract energy into the useful work. The high pressure of the water from the head will be applied to the blade. Long term exposure of this high pressure will lead to the blade metal become fatigue and cause a fracture. The objectives of this study are to obtain the total deformation and stresses that act on the blade. There are few new designs of the blade from to previous studies will be consider. The turbine blade is simulated using the finite element method (FEM) based on the commercial code ANSYS. The result of the datum design is compared with all new designs in order to analyze if the new design is better or vice versa from the datum design. The best design is selected based on the largest percentages difference of the selected design compared to the datum design. Using statistical analysis, paired comparison design is selected to compare between the datum design and the selected design, i.e. do the results have a significant difference. It can be concluded that the new design is better in handling the total deformation, stress and strain.

Keywords Francis turbine · Finite element method · Total deformation · Stress · Strain

K. Shahril (✉) · A. Tajul · M. S. M. Sidik · K. A. Shamsuddin
A. R. Ab-Kadir
Universiti Kuala Lumpur Malaysian Spanish Institute, Kulim Hi-Tech Park,
09000 Kulim, Kedah, Malaysia
e-mail: khairuls@unikl.edu.my

A. Tajul
e-mail: tajuladli@unikl.edu.my

M. S. M. Sidik
e-mail: msabri@unikl.edu.my

K. A. Shamsuddin
e-mail: khairulakmal@unikl.edu.my

A. R. Ab-Kadir
e-mail: ahmadrazlee@unikl.edu.my

© Springer Nature Switzerland AG 2019
M. H. Abu Bakar et al. (eds.), *Progress in Engineering Technology*,
Advanced Structured Materials 119, https://doi.org/10.1007/978-3-030-28505-0_15

1 Introduction

A Francis turbine is mostly used in a hydropower plant due to the great efficiency compared to other turbines such as the Pelton and the Kaplan turbine. It works under a medium head (50 m to 300 m) and is handling a medium quantity of the water. The blade is the most critical part of the turbine, it must withstand high pressure from the water to extract energy into useful work. The proper design of the blade will affect the total deformation, stress and strain, i.e. a good design will increase the lifetime of the blade and requires less maintenance.

The pressure of the water produces stress on the blade in long term can lead to metal fatigue. Metal fatigue is a phenomenon which results in the sudden fracture of a component after a period of cyclic loading in the elastic regime. Failure is the end result of a process involving the initiation and growth of a crack, usually at the site of a stress concentration on the surface. This can be found in the welded areas on the blade. Cavitation is the development of vapor structures in an originally liquid flow due to pressure drops in the flow itself, and the collapse in high pressure regions. This will develop erosion on the blade.

The scope of this study are to simulate the pressure that acts on the micro Francis turbine blade on the datum design and the new design. A new design blade is a design based on the previous study, fillet design, lean design and thicker design. By using FEM simulation, the result of the simulation of the total deformation, stress and strain of the datum design will be compared all new designs. Statistical analysis is applied to verify the difference to know if the new design is better or vice versa from the datum design.

2 Literature Review

D. Frunzverde et al. performed a failure analysis of a Francis turbine runner and found that in the blade operating at partial load causes pressure fluctuation due to vortex rope in draft tube [1]. This problem leads to the strong vibration and noise causing the failure on the blade where fatigue occurs that can lead to a crack as shows in Fig. 1. A fatigue crack tends to occur at the early stage or after a decade of operation. This problem usually occurs on the welded area where the blade is joint with the crown. Furthermore, the welding is done as a T-joint where high pressure from the water head hits the blade. This can be proven by simulation of the real condition using CFD where the geometry is simplified and does not contain the fillet, crown and band, which could affect the results due to stress concentration. The mesh is consisting of 61,140 solid type elements, connected in 38,186 nodes. The imposed boundary conditions were fixed supports on the two ends of the blade (link between blade and crown and band). The result shows that the loading edge has high pressure that is 0.74605 Mpa and lean to the tail edge. The conclusion can be made that the welding technique must be improved [1].

Fig. 1 Fatique crack on blade due to strong vibration

Thapa et al. stated in his article that cavitation occurs on a Francis turbine exposed to sediment erosion [2]. Cavitation is the development of vapor structures in an originally liquid flow due to pressure drops in the flow itself, and the collapse in high pressure regions. According to the Bernoulli equation, at the point where the velocity is at its maximum, the pressure is at its minimum and the risk of cavitation is high.

There are a few suggestions to decrease the erosion such as increase turbine diameter, thicker runner blades, fewer runner blades and lastly coating on exposed parts. There are disadvantages when changes are done to the design that may affect the efficiency and increase the tendency to vibration caused by von Karman vortices, increased surface roughness and lastly increased material costs.

Leading edge cavitation occurs as shown in Fig. 2 at location (A) and (B), Travelling bubble cavitation usually occurs at location (C) and inter-blades cavitation usually occurs at location (D). Von Karman vortex cavitation occurs at the trailing edge, but will damage the trailing edge due to pressure pulsations.

The importance of the pressure distribution is reflected in the blade leaning. The blade leaning angle is given as the angle that is normal to the flow direction, meaning; by tilting the vertical inlet, leaning to the blade is introduced as shown in Fig. 3. It can also be expressed as an angular displacement of each streamline at the inlet. By leaning the blade, the pressure distribution from hub to shroud can be

Fig. 2 Cavitation Locations
A and **B** leading edge
C travelling buble
D inter-blades

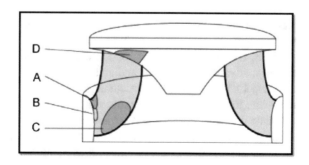

Fig. 3 X-blade (lean) design

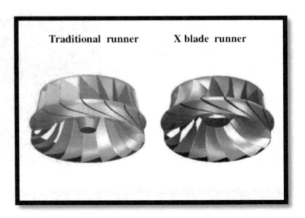

adjusted, hence low pressure zones at hub or shroud can be removed and thereby reducing the cross flow on the blade. The new design of the blade leaning is the X-blade [2].

3 Methodology

3.1 3D Modelling of the Proposed Design

The precise geometry of an actual Francis turbine blade is measured using FlexScan3D of a 3D scanner as shown in Fig. 4. Then it was converted into a 3D model in Catia V5. The Catia V5 file was transfer into SOLIDWORKS2015 in order to reconstruct for smooth surface and removal of any unnecessary parts [3]. Figure 5 shows the datum design after being finalized and that dimension for other design is based on the datum model, but the thickness of the blade, fillet on the joint area and lean design have been changed. The changes are based on a previous study [4].

3.2 ANSYS Simulation

Computer simulation is used due to limitation to solve the behavior of the complex surrounding. In this study ANSYS is selected due the advantages of ANSYS ANSYS, i.e. rich functionality, consistency with theoretical foundation employing the most advanced algorithms and architectural flexibility [5].

For a FEM analysis, the structure is represented as finite elements. These elements are joined at particular points which are called as nodes. The FEA is used to calculate the deflection, stresses, strain or the buckling behavior of the member [5]. A static structural analysis is selected in ANSYS to perform this simulation. All the mechanical properties of the turbine is entered into the engineering data, see Table 1.

Fig. 4 3D scanner

Fig. 5 Datum design

Table 1 Boundary conditions and material properties

Boundary condition	Default value
Fixed support location	On the coupling
Force location	On the blade
Force value	53.852 N
Young's modulus	200 GPa
Poisson's ratio	0.3
Density	7850 kgm^{-3}
Tensile yield strength	2.5E+08 N/m^2

After that, the 3D geometry from SOLIDWORKS is imported into ANSYS. Due to the easier user interface of SOLIDWORKS, all the 3D modeling is done in SOLIDWORKS, and the modeling is imported into the ANSYS. This procedure will save time without geometry modeling in ANSYS. The next step is the meshing of the geometry. The most suitable meshing size is 0.001 m due to the size of the turbine. A smaller size than that will increase the time to mesh and increase the size of the result file without increasing the accuracy. Force location and fixed support location is applied on certain parts as shown in Table 1. For the result, the total deformation, von mises stress and von mises strain are evaluated.

3.3   Paired Comparison Design

By using Minitab 17, paired comparison design can be done in a few minutes compared to manual human calculation. The best design is selected based on the highest percentage difference in total deformation, stress and strain. Paired comparison design is used to compare the datum design and the selected design. All the values from the result is entered into the Minitab 17.

4   Results and Discussion

4.1   Analysis Graph Result

Figure 6 shows the result of the total deformation for all design configurations, whereas the total deformation for the datum design is 1.22E−05 mm. For this single parameter, the result shows a lower significant difference as compared to a combination of parameters. The total deformation for the lean design is

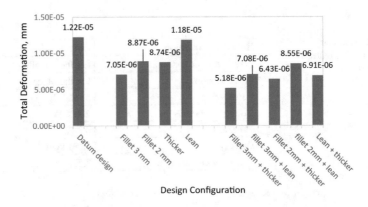

Fig. 6 Comparison of total deformations

11.831E−06 mm. The difference is 3.31% from the datum design but is vice versa for the total deformation for fillet 3 mm design is 7.05E−06 mm and the difference almost 42.35% from the datum design. For the fillet 2 mm and thicker design, the total deformation is almost the same, i.e. 8.87E−06 mm and 8.74E−06 mm and the difference is 27.49 and 28.59%, respectively.

For the combination parameter, the most significant difference is for the fillet 3 mm with thicker that is 6.91E−05 mm with a difference of almost 57.63%. This result is followed by the fillet 2 mm with thicker, the total deformation is 6.43E−06 mm and the difference is 47.46%. Total deformation for the lean with thicker is 6.91E−06 mm and the difference is 43.56% Fillet 3 mm with lean total deformation is 7.08E−06 mm and the difference is 42.11%. Lastly the total deformation fillet 2 mm with lean is 8.55E−06 with different 30.12%.

The largest difference of 57.69% to the datum design is obtained for fillet 3 mm with thicker. This design proposal is most suitable in handling the total deformation, where the used material is a steel allow that has a low strain value to prevent the material to elongate until it reaches the maximum point that can break the material.

Figure 7 shows the result of stress for all design configurations, whereas the stress for the datum design is 4.35E+07 N/m^2. The poorest in handling stress is the lean design. The stress is 3.46E+07 N/m^2 and the difference with the datum design is 20.44% followed by fillet 2 mm, whereas the stress is 3.37E+07 N/m^2 with the different is 22.64%. Stress for thicker design is 2.95E+07 N/m^2 and the different is 32.25%. The best design in handling the stress is the fillet 3 mm where the stress is 2.87 + 07 N/m^2 and the difference is 34.02%.

For the combination parameter, the best configuration in handling this parameter is the lean with thicker where the stress is only 1.62E+07 N/m^2 and a difference of almost 62.87% from the datum design. The stress for the filled 3 mm with thicker is 2.11E+07 N/m^2 and the different is 51.58% more than half of the datum design.

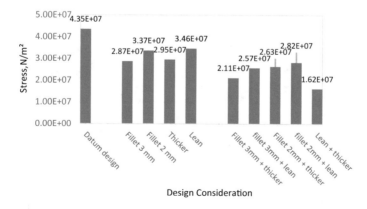

Fig. 7 Comparison of stress values

There are almost the results same for the 3 mm with lean and the 2 mm with thicker where the stress is 2.57+07 N/m^2 and 2.53E+07 N/m^2 and the difference with the datum configuration is 41.05 and 39.51% respectively. Lastly for combination parameter, the worst in handling the stress is the fillet with 2 mm with lean. The stress is 2.82E+07 N/m^2 and the difference with the datum configuration is 35.19%.

The largest difference with datum design is the fillet lean with thicker more than half that is 62.87% most suitable in handling the stress, the fillet area on the joining area distributed the pressure evenly and lower the stress concentration. Figure 8 shows the result of strain for all designs, whereas the strain for the datum strain for datum design is 2.18E+07. For the single parameter, t the worst design is the lean design where the strain is 1.73E−04, where the different with the datum design is 20.55%. It is followed by the fillet 2 mm design where the strain is 1.70E−04 and the difference is 22.22%. For the thicker design and the fillet 3 mm the strain is almost the same, where the strain is 1.48E−04 and 1.44E−04 and the difference is 32.09 and 34.08%, respectively.

For the combination parameter, the greatest strain is for the combination lean with thicker, where the strain is 8.22E−05 and the different is almost 62.27% from the datum design. The strain for fillet 3 mm with thicker is 1.05E−04 and the difference is 51.61%. For the fillet 3 mm with lean and fillet 2 mm with thicker the strain is almost the strain is 1.28E−04 and 1.32E−04 and the difference is 41.14 and 39.55%, respectively. The worst in strain value is for the fillet 2 mm with lean. The strain is 1.41E−04 and the difference is 35.29% with the datum design.

The largest different with the datum design is the lean with thinker, the difference with the datum design 62.27%. Strain the ratio of the representing length with the original length, in this case, a lower strain is better and this is due to the fact the material.

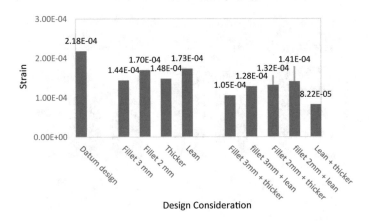

Fig. 8 Comparison of strain values

4.2 Paired Comparison Design

From the result, the graph shows that the lean with thicker performs good in regards to stress and strain although the total deformation is not the best. The lean with thicker design is still good from the datum design and the difference with the datum design 43.56%, i.e. almost half from the datum design. The lean with thicker design is selected for best design in this thesis. To prove that there are significant differences with the datum design for the lean with thicker design, the paired comparison is done for the total deformation, stress and strain result.

Figure 9 shows the result of the paired comparison design for the total deformation between the datum designs with the lean with thicker design. For the significant level 5%, the P-value is lower than the significant level, it means there are significant differences for both design.

Figure 10 shows the result of the paired comparison design for the stress between the datum designs with the lean with thicker design. For the significant level 5%, the P-value is lower than the significant level, it means that there are significant differences for both design.

```
Paired T-Test and CI: total deformation for datum, total deformation for selected

Paired T for total deformation for datum - total deformation for selected

                         N      Mean      StDev     SE Mean
total deformation for da 10  0.000006  0.000004  0.000001
total deformation for se 10  0.000003  0.000002  0.000001
Difference               10  0.000003  0.000002  0.000001


95% CI for mean difference: (0.000001, 0.000004)
T-Test of mean difference = 0 (vs ≠ 0): T-Value = 4.70  P-Value = 0.001
```

Fig. 9 Paired comparison design for total deformation

```
Paired T-Test and CI: stress for datum, stress for selected

Paired T for stress for datum - stress for selected

                      N      Mean      StDev    SE Mean
stress for datum     10  21711421  14670280  4639150
stress for selected  10   8099845   5412428  1711560
Difference           10  13611576   9258440  2927776


95% CI for mean difference: (6988487, 20234665)
T-Test of mean difference = 0 (vs ≠ 0): T-Value = 4.65  P-Value = 0.001
```

Fig. 10 Paired comparison design for stress

Paired T-Test and CI: strain for datum, strain for selected

```
Paired T for strain for datum - strain for selected

                          N      Mean       StDev     SE Mean
strain for datum          10   0.000109    0.000073   0.000023
strain for selected       10   0.000041    0.000028   0.000009
Difference                10   0.000068    0.000046   0.000014


95% CI for mean difference: (0.000035, 0.000101)
T-Test of mean difference = 0 (vs ≠ 0): T-Value = 4.70  P-Value = 0.001
```

Fig. 11 Paired comparison design for strain

Figure 11 shows the result of the paired comparison design for the strain between the datum designs with the lean with thicker design. For the significant level 5%, the P-value is lower than the significant level, it means that there are significant differences for both design.

5 Conclusions

The selected design is the lean with thicker due to good performance in stress and strain although not the best in regards to total deformation. For the total deformation, the difference with the datum design is 43.56%. For stress, the difference with the datum design is 62.87% and last but not least, the different with the datum design is 62.27%. This is proven by statistical analysis, paired comparison design for total deformation, stress and strain, The paired comparison design shows that there are significant differences for all of three result.

Acknowledgements The authors gratefully acknowledge financial support for this work by the Universiti Kuala Lumpur Malaysian Spanish Institute (UniKL MSI).

References

1. Frunzăverde, D., et al.: Failure analysis of a Francis turbine runner. IOP Conf. Ser. Earth Environ. Sci. **12**, 012115 (2010)
2. Singh, B., Thapa, B., Gunnar, O.: Current research in hydraulic turbines for handling sediments. Energy **47**(1), 62–69 (2012)
3. Gómez-López, L.M., Miguel, V., Martínez, A., Coello, J., Calatayud, A.: Simulation and modeling of single point incremental forming processes within a solidworks environment. Procedia Eng. **63**(2005), 632–641 (2013)

4. Bhashyam, G.R.: ANSYS Mechanical—A Powerful Nonlinear Simulation Tool, no. September, 2002
5. Dharmadhikari, S.R., Mahakalkar, S.G., Giri, J.P., Khutafale, N.D.: 'Design and analysis of composite drive shaft using ANSYS and genetic algorithm' a critical review. Int. J. Mod. Eng. Res. **3**(1), 490–496 (2013)

Deep Contractive Autoencoder-Based Anomaly Detection for In-Vehicle Controller Area Network (CAN)

Siti Farhana Lokman, Abu Talib Othman, Shahrulniza Musa
and Muhamad Husaini Abu Bakar

Abstract With the emerging wireless technology integrated into modern vehicles, this introduces an enormous number of vulnerabilities for adversaries to compromise the vehicle internal system. Nonetheless, the attacks can be alleviated using anomaly detection mechanism which have been proven to be effective in monitoring and detecting attacks. In this paper, we developed an anomaly detection using an unsupervised deep learning-based approach, known as Deep Contractive Autoencoders (DCAEs). The DCAEs, which is one of the regularize autoencoders model imposed a different penalty term to the CAN data representation in order to encourage robustness towards small changes. To accomplish this purpose, we captured CAN bus data from three different vehicles, pre-processed them using the max absolute normalization, and evaluated the model over three types of attacks. Finally, the experimental results demonstrated that DCAEs yield a 91–100% detection rate which outperformed other variants of regularized autoencoders.

Keywords Controller area network · Unsupervised learning · Anomaly detection · Autoencoder · Deep neural network

S. F. Lokman (✉) · A. T. Othman · M. H. Abu Bakar
System Engineering and Energy Laboratory, Universiti Kuala Lumpur Malaysian Spanish
Institute, Kulim Hi-Tech Park, 09000 Kulim, Kedah, Malaysia
e-mail: farhana.lokman@s.unikl.edu.my

A. T. Othman
e-mail: abutalib@unikl.edu.my

M. H. Abu Bakar
e-mail: muhamadhusaini@unikl.edu.my

S. Musa
Universiti Kuala Lumpur, Malaysian Institute of Information Technology, Vision City,
Bandar Wawasan, 50300 Kuala Lumpur, Malaysia
e-mail: shahrulniza@unikl.edu.my

© Springer Nature Switzerland AG 2019
M. H. Abu Bakar et al. (eds.), *Progress in Engineering Technology*,
Advanced Structured Materials 119, https://doi.org/10.1007/978-3-030-28505-0_16

1 Introduction

These days, modern vehicles are no longer mechanical- or hydraulic-based, instead, they are mostly being replaced by a number of electronic devices which are called electronic control units (ECUs). These ECUs exchange messages among other electronic components such as sensors and actuators through one of the most common automotive communication bus system, the controller area network (CAN) [1]. This CAN bus system is responsible to ensure all critical parts of vehicles like the engine, braking, steering, and other safety systems are performed accordingly. The communication that takes place between ECUs on the CAN bus also is connected to other communication systems such as LIN, FlexRay and MOST. Figure 1 shows the overall communication system in most modern vehicles.

Apart from that, some of the ECUs that are equipped with wireless technology allows them to communicate with the outside world. Consequently, the interaction between the internal vehicle system with the external environment introduces an array of vulnerabilities which could allow attackers to have legitimate access into one bus system, and eventually compromise the entire vehicle communication system. Judging from this fact, a great amount of effort has been put in by security researchers to demonstrate the attack surface that exists in modern vehicles. Checkoway et al. explored a broad array of remote exploitation through Bluetooth, mechanic tools, cellular radio, CD players, and further, another attack vector like wireless technology also can be feasible to perform long-range attacks [3]. The adversaries can leverage the compromised ECUs by flooding the CAN bus with a large number of CAN packets and eventually take over the entire car system [4].

To cope with the challenges illustrated above, there have been several proposals found in the literature trying to develop security detection mechanisms in detecting malicious activities in the CAN bus system. The artificial neural network (ANN)-based anomaly detection is recently one of the emerging approaches that have been

Fig. 1 In-vehicle
communication network
illustrated by Li [2]

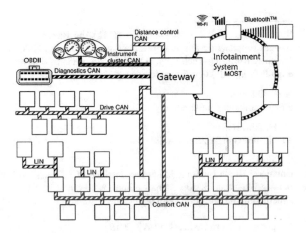

employed in CAN bus networks [5–7]. The advantages of this approach that makes CAN bus favors this approach are because its capability to generalize and adapt to changes in the domain, and finally being able to cope with noise [8].

Wasicek and Weimerskirch [7] were the first ones that introduced the ANN-based anomaly detection for CAN bus networks. They proposed a bottleneck ANN with backpropagation architecture in classifying normal engine behavior as a baseline detection feature i.e., torque, speed, and revolutions per minute (RPM). However, the result depicted an inclination towards no difference between attack and normal data. Kang et al. [5] developed a semi-supervised deep neural network (DNN) on three CAN bus IDs that were generated using the simulator. Besides manipulating CAN data to deceive the proposed system, the author also added natural randomness using Gaussian noise to the value information of CAN data. Farhana Lokman et al. [9] implemented an unsupervised stacked sparsed autoencoders (SSAEs) anomaly detection as acquiring a completely labeled data from a real-time CAN bus is inconvenient. However, the proposed method was not specifically evaluated on each type of attack and the overall performance yielded a low ROC result. By conducting an experiment on a different kind of attack can determine which attack influences the overall anomaly detection performance while at the same time improving the structure to reassure efficiency and robustness of the model.

Hence, the proposed solution introduces a novel approach which is based on the deep contractive autoencoders (DCAEs) [10] in detecting the malicious pattern in CAN bus data. Precisely, the DCAEs model is one of the regularized autoencoders variants like SSAEs and denoising autoencoders (DAEs) [11]. The idea of DCAEs is to ensure the robustness of learned CAN bus data representation to small changes, noise and missing inputs in the training samples. To achieve this purpose, the DCAEs imposed a different penalty term in the lost function to the representation. As a result, the DCAEs model surpasses results attained by other regularized autoencoders.

The remaining of this paper is organized into three sections: Sect. 2 describes our proposed DCAEs method and presents the CAN bus dataset used for experimental purposes. Section 3 explains the overall results obtained from the experiment. Finally, Sect. 4 concludes this paper along with future work.

2 Methodology

2.1 The DCAEs Design

As mentioned earlier, the DCAEs design [10] develops an explicit regularizer or penalty term on the latent representation $h = f(x)$, reducing the derivative of f to be smaller as possible. The steps taken by the Gradient descent (GD) will be large if the loss function of the derivative is high, and will take a small step if the loss function of the derivative is low. Thus, to achieve an optimal value, it should have a very small number (precisely, 0) of the derivative. The penalty term is shown below. The delta of λ indicates a scaling factor:

$$\Omega(h) \;=\; \lambda \left\|\frac{\delta f(x)}{\delta(x)}\right\|_f^2 \tag{1}$$

The $\Omega(h)$ denotes the Frobenius norm of the Jacobian matrix, which is the sum squared of all vectors in the matrix associated with the encoder activation sequence, with respect to CAN data input x. Finally, to construct multiple hidden layers of DCAEs, we stacked autoencoders one after another. Previous work [12] suggested that having a deep layer of the network to compress data representation could improve the model performance.

2.2 Experimental Setup

The methodology of the learning procedure is illustrated in Fig. 3 In this section, we described a different approach performed in each step.

(i) CAN bus dataset

The CAN bus dataset that we used in this paper is logged from three different vehicles; Toyota Camry, Hyundai Elantra, and Perodua Axia. During the data acquisition, we used the DB9 cable which was plugged into an OBD-II port. Most vehicles that have a CAN bus system are equipped with the OBD-II port around the steering wheel, allowing manufacturers or security researchers to diagnose or capture CAN bus data. We connected the cable with the CAN bus data sniffer device in order to allow communication between the laptop and the CAN bus system. We captured the CAN bus data approximately for 30 minutes of driving, which also involves idling, speeding, braking, parking, locking doors, turning the lights on and off. The process of acquiring CAN bus data from one of the vehicles is depicted in Fig. 2.

The captured CAN bus data is pre-processed first before training them. The pre-processing phase is comprised of three steps. (1) Feature reductions, where we exclude unnecessary attributes in CAN data payload such as timestamp and size. (2) Features conversion, where we converted the hexadecimal data into decimals since the DNN-based model could not process string data. (3) Features normalization, we scaled the data using max absolute normalization to ensure their scales are equal (Fig. 3).

To evaluate the model, we injected the CAN bus traffic with different kinds of attacks described below:

Fig. 2 The process of CAN bus data acquisition on Perodua Axia

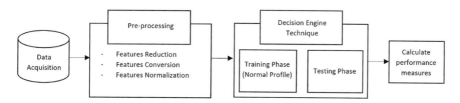

Fig. 3 The methodology of the learning procedure

(a) DoS attack: the aim of this attack is to gain a dominant state in the CAN bus traffic. We constantly flooded the traffic with the highest priority data containing CAN ID and Data field values of '0x000'. This DoS attack will make other legitimate CAN data to back off from the normal state of traffic.

(b) Fuzzy attack: we manipulated the normal CAN ID and Data fields with a completely unique and random value. This attack is aimed to make the system confused.

(c) Impersonate attack: we manipulated the specific vehicle components that are associated with RPM and gear with a constant value; '0xfff' and '0x000'. This attack is aimed to make legitimate RPM and gear components to behave abnormally.

All attacks above are performed at a short period of time which ranges between 1 and 4 milliseconds. Next, we divided the normal and injected CAN bus data into training and testing dataset (Table 1).

Table 1 The CAN bus datasets used during the experiment

Vehicle	Features		Attack types	Training packets	Test packets	
	Unique ID	Unique data		Normal packets	Normal packets	Injected packets
Toyota Camry	103	259,748	DoS	1,000,000	600,000	300,000
			Impersonate	1,000,000	600,000	300,000
			Fuzzy	1,000,000	600,000	300,000
Hyundai Elentra	70	48,931	DoS	1,000,000	600,000	300,000
			Impersonate	1,000,000	600,000	300,000
			Fuzzy	1,000,000	600,000	300,000
Perodua Axia	26	259,748	DoS	1,000,000	600,000	300,000
			Impersonate	1,000,000	600,000	300,000
			Fuzzy	1,000,000	600,000	300,000

(ii) Experimental Environment and Performance Measure

We compared the efficiency and robustness of the detection rate based on the receiver operating characteristic (ROC) performance measure; where it plots the true positive rate (TPR) against the false positive rate (FPR). It offers a way in making comparisons between methods since its discrimination threshold settings are varied. The most optimal discriminator is a 100% detection with zero FPR. We also examined the usefulness of two different types of error metrics in comparing between the actual data and the predicted data: mean squared error (MSE) [13] and mean absolute error (MAE) [14]. Generally, MSE calculates the average of the squared difference between actual values and prediction errors, while MAE measures an average absolute difference between actual values and predictions. The mathematical model for both error metrics is shown below:

$$MSE = \frac{1}{N}\sum_{i=1}^{N}(y_i - {}' y_i)$$ (2)

$$MAE = \frac{1}{N}\sum_{i=1}^{N}|y_i - {}' y_i|$$ (3)

Lastly, we compared our proposed DCAEs model with other regularized autoencoder variants in determining the best anomaly detection solution for CAN bus data:

(a) stacked sparsed autoencoders with *l1* regularizer constraints (SSAEs);
(b) stacked denoising autoencoders (SDAEs).

3 Results and Discussion

The qualitative experimental results using DCAEs are exhibited in Table 2. The rows that are highlighted with bold fonts determined the best ROC in detecting three different types of attacks. While the underlined and italic font signified the highest detection rate using different error metrics (Fig. 4).

From the experimental results presented above, some initial observations can be concluded. Our proposed DCAEs anomaly detection model obtained the highest detection rate which ranges between 91 to 100% and performed consistently on every attack compared to other regularized autoencoders approaches. We further described and evaluated different detection models in three attack scenarios below:

(i) DoS attack

It can be observed that the performance of the SSAEs and DAEs models is degraded when tested on DoS attack, however, the DCAEs detection rate is still

Table 2 ROC result among three different regularization autoencoders

Car model	Attack type	SSAE		SDAE		DCAE	
		MAE (%)	MSE (%)	MAE (%)	MSE (%)	MAE (%)	MSE (%)
Perodua Axia	DoS	94.70	94.70	80.70	80.30	**100.00**	**100.00**
Hyundai Elentra		63.70	56.20	**76.30**	77.10	75.80	75.00
Toyota Camry		26.90	26.90	84.10	83.60	**98.30**	98.10
Average		_61.77_	59.27	_80.37_	80.33	_91.37_	91.03
Perodua Axia	Fuzzy	81.30	83.20	81.90	83.20	**94.70**	**94.70**
Hyundai Elentra		59.40	63.00	43.40	46.50	90.60	**91.30**
Toyota Camry		79.00	81.90	77.00	75.20	**91.90**	**92.40**
Average		73.23	_76.03_	67.43	_68.30_	92.40	_92.80_
Perodua Axia	Impersonate (Brake)	**100.00**	**100.00**	99.30	99.30	**100.00**	**100.00**
Hyundai Elentra		97.50	99.50	**100.00**	**100.00**	**100.00**	**100.00**
Toyota Camry		96.40	96.40	**100.00**	99.40	98.30	98.50
Toyota Camry		96.40	**96.40**	**100.00**	99.40	98.30	**98.50**
Average		97.97	_98.63_	_99.77_	99.57	99.43	_99.50_

much higher than SSAEs and DAEs. The model system might get confused when trying to discriminate between the actual data and the injected data because several normal CAN features contained similar values like a DoS attack. It can be noted that, since the variances of the CAN feature with the attack instances are close, the SSAEs and DAEs performance declining when it comes to decision making.

(ii) Fuzzy attack

Although the DCAEs performance maintains the highest ROC score in all vehicle datasets, however, the percentage of the detection rate drops approximately from 4 to 9% compared to the previous attack. Similarly, to SSAEs and DAEs. This might be due to the variations of the fuzzy attack data that has slightly similar values between normal and abnormal instances with the actual data. This indicated the complete randomness and complicated patterns of fuzzy attacks that nearly imitate normal patterns.

(a)

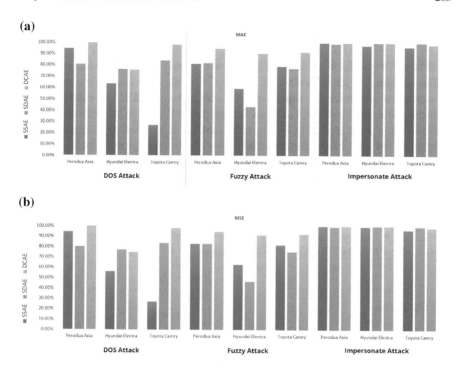

(b)

Fig. 4 The ROC results for three attack scenarios; DoS, fuzzy and impersonate **a** figure presents MSE error metric result **b** figure shows MAE error metric result

(iii)   Impersonate attack

In contrast with other attack scenarios, all the detection model's performance, in this case, are encouraging. Since the injected impersonate attack contained constant values that do not exist in normal CAN data, thus explains the results of no false positive error.

In terms of error metrics used in this experiment, both MSE and MAE produced slightly similar results on most detection model. However, the MSE error metric is prone to performed much better on the fuzzy attack, while MAE performed well on DoS and impersonate attack. Thus, it can be concluded that the MSE error metric is robust when dealing with random values attack scenario while MAE error metric is effective when dealing with constant values attack scenario.

4 Conclusions

In this paper, we have utilized an unsupervised deep learning model in discriminating different types of anomalies within CAN bus data using the prior normal instances to construct the representations. Even though DCAEs-based anomaly detection describes the traffic of CAN bus system well, however, there still remains a challenging point in distinguishing malicious traffic resulted by 'normal faulty of electronic devices' from malicious traffic resulted by 'intentional attacks incurs by adversaries'. Nevertheless, the DCAEs-based anomaly detection is still robust towards lack of 'identified attack behaviors for in-vehicle CAN bus', due to more attack patterns and framework for vehicles have become subsequently discovered.

Acknowledgments This study was supported by the Short-Term Research Grant (grant number str17001), funded by the Center for Research and Innovation, Universiti of Kuala Lumpur and System Engineering and Energy Laboratory (SEELab).

References

1. Miller, C., Valasek, C.: A survey of remote automotive attack surfaces. Black hat USA (2014)
2. Li, J.: CANsee-an automobile intrusion detection system. In: Presentation Slides on Hack In The Box Security Conference (HITBSecConf) (2016)
3. Checkoway, S., McCoy, D., Kantor, B., Anderson, D., Shacham, H., Savage, S.,... Kohno, T.: Comprehensive experimental analyses of automotive attack surfaces. In: USENIX Security Symposium vol. 4, (2011, August)
4. Koscher, K., Czeskis, A., Roesner, F., Patel, S., Kohno, T., Checkoway, S.,... Savage, S.: Experimental security analysis of a modern automobile. In: 2010 IEEE Symposium on Security and Privacy, pp. 447–462. IEEE (2010, May)
5. Kang, M.J., Kang, J.W.: Intrusion detection system using deep neural network for in-vehicle network security. PLoS ONE **11**(6), e0155781 (2016)
6. Taylor, A., Leblanc, S., Japkowicz, N.: Anomaly detection in automobile control network data with long short-term memory networks. In: 2016 IEEE International Conference on Data Science and Advanced Analytics (DSAA), pp. 130–139. IEEE (2016, October)
7. Wasicek, A., Weimerskirch, A.: Recognizing manipulated electronic control units. No. 2015–01-0202. SAE Technical Paper (2015)
8. Lokman, S.F., Othman, A.T., Abu-Bakar, M.H.: Intrusion detection system for automotive Controller Area Network (CAN) bus system: a review. EURASIP J. Wirel, Commun. Netw. **2019**(1), 184 (2019)
9. Farhana Lokman, S., Talib Othman, A., Husaini Abu Bakar, M., Razuwan, R.: Stacked sparse autoencoders-based outlier discovery for in-vehicle controller area network (CAN). Int. J. Eng. & Technol., **7**(4.33), 375–380 (2018). http://dx.doi.org/10.14419/ijet.v7i4.33.26078
10. Rifai, S., Vincent, P., Muller, X., Glorot, X., Bengio, Y.: Contractive auto-encoders: explicit invariance during feature extraction. In: Proceedings of the 28th International Conference on International Conference on Machine Learning, pp. 833–840. Omnipress (2011, June)
11. Vincent, P., Larochelle, H., Bengio, Y., Manzagol, P.A.: Extracting and composing robust features with denoising autoencoders. In: Proceedings of the 25th International Conference on Machine Learning, pp. 1096–1103. ACM (2008, July)

12. An, J., Cho, S.: Variational autoencoder based anomaly detection using reconstruction probability. Spec Lect IE **2**, 1–18 (2015)
13. Lehmann, E.L., Casella, G.: Theory of Point Estimation. Springer Science & Business Media (2006)
14. Willmott, C.J., Matsuura, K.: Advantages of the mean absolute error (MAE) over the root mean square error (RMSE) in assessing average model performance. Climate Res. **30**(1), 79–82 (2005)

Design and Temperature Analysis of an Aluminum-Air Battery Casing for Electric Vehicles

Mohamad Naufal Mohamad Zaini, Mohamad-Syafiq Mohd-Kamal,
Mohamad Sabri Mohamad Sidik and Muhamad Husaini Abu Bakar

Abstract The aluminum-air battery receives more attention to applications in electronic mobile devices, transportation systems, and has a higher energy density than other metal-air batteries. However, the aluminum-air battery is still not widely commercialized due to unacceptably thermal issues. Hence, this study focuses on the development of an aluminum-air battery casing, studies the performance of the aluminum-air battery and thermal distribution analysis by using thermography. A single cell with dimensions of 10 cm × 10 cm × 3 cm with an anode area of 6.5 cm^2 and an air cathode area of 6.5 cm^2 is designed. In addition, 1 M of NaOH acts as the electrolyte of the battery. The aluminum-air battery temperature distribution is determined by a thermal imaging camera. The maximum temperature of 34 °C has been found as the reaction occurs. The result of the battery tests shows that the battery can produce a maximum voltage of 1.5 V and has a constant current value of 40 mA. The discharge rate of the battery indicates that one cell can operate for 10 h. Thus, the proposed design for the battery casing has functioned at the optimal condition.

Keywords Aluminum-Air battery · Aluminum anode · Metal-Air battery · Thermal analysis · Electric vehicle

M. N. Mohamad Zaini · M.-S. Mohd-Kamal · M. S. Mohamad Sidik · M. H. Abu Bakar (✉)
System Engineering and Energy Laboratory, University Kuala Lumpur Malaysian Spanish
Institute, Kulim Hi-Tech Park, 09000 Kulim, Kedah, Malaysia
e-mail: muhamadhusaini@unikl.edu.my

M. N. Mohamad Zaini
e-mail: mr.naufal14@gmail.com

M.-S. Mohd-Kamal
e-mail: msyafiq.kamal@s.unikl.edu.my

M. S. Mohamad Sidik
e-mail: msabri@unikl.edu.my

© Springer Nature Switzerland AG 2019
M. H. Abu Bakar et al. (eds.), *Progress in Engineering Technology*,
Advanced Structured Materials 119, https://doi.org/10.1007/978-3-030-28505-0_17

1 Introduction

Referring to the new lifestyle nowadays, fuel consumption became a critical issue in selecting the car in daily life. Hence, the suitable vehicle that fits a human need is an electric vehicles (EV). Besides, there are a variety of technologies available in EV's that have different capabilities that can accommodate different driver's needs. The usage of batteries for high power portability has spread in various applications. One of the green technologies that is popular nowadays is an electric vehicle. EV's have existed for more than a century by now. In 1899, a Belgian electric vehicle powered by a lead-acid battery was able to reach 30 m/s [1]. However, the lack of progress in batteries hindered the development of EVs and it was not until recently that electric and hybrid vehicles re-emerged. EVs are powered entirely by electric propulsion systems, while hybrid vehicles have two or more power sources—normally an ICE coupled to an electric motor/generator powered by an electric energy storage system.

In the automotive field, usually, the batteries are used for the starter and now the usage of batteries is getting essential as the power supplier for an electrical vehicle [13]. The fabrication of a battery involves the anode and cathode as electrode and electrolyte. The type of electrode and electrolyte can be different depending on its purpose. Each unique combination can provide a different value of voltage [3]. Therefore, an alternative strategy is desired to develop novel energy storage and conversion systems with sufficient theoretical energy density required for future applications [11]. Among these new energy storage systems, metal-air batteries have gained great interest due to their high energy density and capacity, low cost (depending on the metal anode), the negligible dependence of their capacity on operating load and temperature, and the constant discharge voltage [10, 12].

Up to now, several different types of metal-air batteries, such as lithium (Li)-air, sodium (Na)-air, potassium (K)-air, zinc (Zn)-air, magnesium (Mg)-air, and aluminum (Al)-air batteries have been extensively studied [2]. Metal-air batteries exhibit high theoretical energy densities ranging between 2 and 10 folds higher than that of state-of-the-art lithium-ion batteries [4, 7]. Metal-air batteries are equipped with a metal anode and an air-breathing cathode through a suitable electrolyte. Due to the open battery configuration of metal-air batteries, the oxygen reagent can be directly received from the surrounding air instead of prior incorporation, thus contributing to their very high theoretical energy densities [8, 9].

In this project, the current issue is that the lithium resource is limited and getting lesser. Therefore, the technologist came out with the idea of the aluminum-air battery. The present aluminum-air battery is the best contender to replace lithium-air battery as aluminum is less expensive, lightweight, abundance in nature and has a higher energy density compared to other metal-air batteries [5, 6]. The importance of this research is to reduce the cost, weight meanwhile increases the power efficiency.

2 Methodology

2.1 Design and Fabrication Process of the Aluminum-Air Battery Casing

The aluminum-air (AA) cell casing shown in Fig. 1 consists of a body that was made from polypropylene with oxygen slot (1). The casing was locked using epoxy glue with hex screw bolt and nut (2). The alkaline electrolyte is poured through the battery cell slot (3) and the volume of the electrolyte for each cell is 0.03 L. The perspex was mounted to the body (4) with a sealing gasket glued on both sides of the battery to see the reaction of the electrode and electrolyte more clearly. The electrolyte-gas mixture exits the cell through the upper outlet. A 100 mm × 65 mm, 1 mm thick aluminum anode (5) is installed between a pair of electrically connected air cathodes (6) with the dimensions of 100 mm × 65 mm each. The total working surface area of the anode-cathodes assembly is 130 cm^2. The anode and cathode are clipped using crocodile clips to hold in place as a current collector (7).

2.2 Experimental Setup

In this analysis, the battery was developed by using several components. To make the single cell aluminum-air battery, one aluminum plate (65 cm^2) with a thickness of 1 mm was used as the anode cell and two carbon meshes (65 cm^2). The device that was used to analyze the battery was a PLX-DAQ battery tester with a 10 W ceramic cement resistor from System Engineering and Energy Laboratory and a 1.5 V DC motor with blade that act as the load that was applied to the battery.

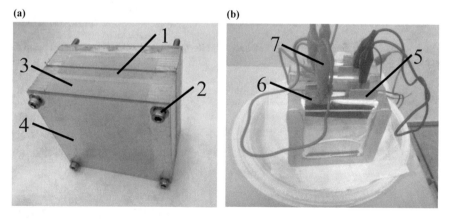

Fig. 1 **a** A aluminum-air battery casing: 1—Oxygen slot; 2—locking device; 3—battery cell slot; 4—perspex. **b** Aluminum-air 4 cell battery: 5—crocodile clip; 6—Aluminum anode; 7—Air cathode

Meanwhile, a multimeter was used to check voltage, current, and resistance of the battery.

The solution for the solid state electrolytes was prepared first. The molar weight of the concentration needs to be calculated to get the exact weight before to dilute in distilled water. The weight of sodium hydroxide (NaOH) needs to be measured. The solution was prepared by mixing the sodium hydroxide (NaOH) that was mixed with distilled water.

2.3   Single Cell and 4 Cell Battery Performance Analysis

The analysis and observation were done by following the step on the schematic diagram in Fig. 2a. The single cell aluminum air battery was connected to the device PLX-DAQ battery tester with 10 W ceramic cement resistor and the device was connected directly to the computer. The experiment was setup as shown in Fig. 2b according to the schematic diagram while the voltage cut off limit was set to 0.12 V.

Meanwhile, for the 4 cell battery testing, the analysis and observation were done by following the step on the schematic diagram in Fig. 3a. The 4 cell aluminium-air battery was connected to the device PLX-DAQ battery tester with 10 W ceramic

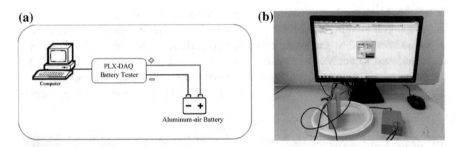

Fig. 2   a Schematic diagram of single cell battery performance analysis **b** Experimental setup for single cell analysis

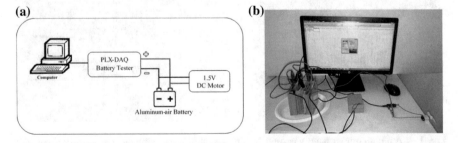

Fig. 3   a Schematic diagram of 4 cell battery performance analysis; **b** Experimental setup for 4 cell analysis

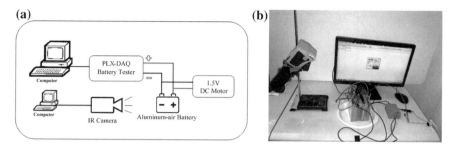

Fig. 4 **a** Schematic diagram of 4 cell battery thermal distribution analysis; **b** Experimental setup for 4 cell thermal distribution analysis

cement resistor and 1.5 V DC motor to give more load on the battery. The device was connected directly to the computer and the experiment was setup as shown in Fig. 3b according to the schematic diagram while the voltage cut off limit was set to 0.12 V.

2.4 Temperature Distribution Analysis

For the temperature distribution analysis, the battery was initially at a normal operating temperature of 28 °C. The analysis process starts with the thermal imager that will record a video of the battery discharge. The recorded video will be analyzed using the Keysight TrueIR U5855A device Analysis and Reporting Tool software to capture the temperature with a thermal imager as shown in Fig. 4. With the enhanced Keysight TrueIR Analysis and Reporting Tool software are able to stream a live thermal image from the computer. This is a very useful tool to monitor the experiment, and perform equipment failure analysis.

3 Discussion

3.1 Single Cell Battery Performance

Based on the graph obtained in Fig. 5, the data generally shows that a single cell aluminum-air battery achieves 1.51 V and 100 mA at the beginning until it stops at 0.12 V cut-off that has been set. Other than that, the results indicate that the lifetime of a single cell aluminum-air battery can be operated for 9 h after 10 W ceramic cement resistor load has been applied. Meanwhile, in Fig. 6, the graph obtained shows that the discharge current of a single cell aluminum-air battery is 100 mAh with 1.51 V at the beginning and drops slowly until the end which is 10 mAh with 0.12 V as shown. It shows that both graphs for the single cell aluminum-air battery

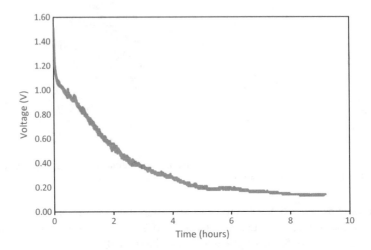

Fig. 5 Graph of voltage over time after 9 h

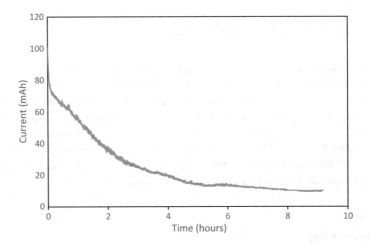

Fig. 6 Graph voltage over discharge capacity

decrease inversely over time. It is decreasing slowly and the energy maintains for a long period of time.

3.2 Battery Performance

From the graph in Fig. 7, the aluminum-air battery was analyzed by using 4 cell battery and the energy capacity achieves 3.88 V at the beginning until it stops at

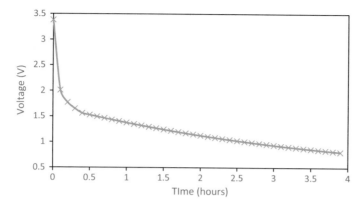

Fig. 7 Graph voltage over time within 4 h

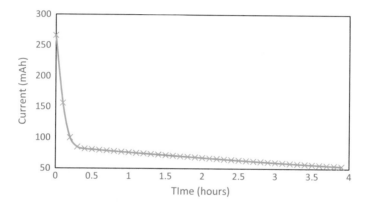

Fig. 8 Graph voltage over discharge capacity

0.5 V voltage cut-off that has been set. Besides, the result shows that the lifetime of the 4 cell aluminum-air battery has been operated for 4 h after a load of 10 W ceramic cement resistor on the battery tester and the 1.45 W DC motor has been applied. Meanwhile, Fig. 8 shows that the discharge current of the aluminum-air battery is 226 mAh at the beginning and drops slowly until the end which is 50 mAh as shown in Fig. 8. It shows that both graphs for the battery decrease inversely over time. It is decreasing slowly and the energy maintains for a long period of time.

3.3 Thermography Test

The experimental setup for the temperature distribution has been analyzing by using 4 cell aluminum-air battery with 10 W ceramic cement resistor as the load on the battery tester and 1.45 W DC motor load were applied on it. The experiment was conducted at 28 °C room temperature. The red focus point shows the maximum temperature on the aluminum-air battery while the white focus point represents the minimum temperature value. The temperature distribution of the aluminum-air battery increases and then decreases as a result occurrence of the parasitic reaction. The process keeps going repeatedly with time which represents the oscillation of a half sine wave graph. During the experiment, the battery temperature starts to increase obviously when reaching 30 °C temperature until it achieves the maximum temperature which is 34 °C.

Based on the temperature versus time graph obtained in Fig. 9a, the temperature starts to increase after 10 s and it keeps increasing gradually until it achieves the maximum temperature of 34 °C after 2 min. Meanwhile, for the 10 min time taken graph in Fig. 9e, the temperature starts increasing gradually until it achieves the maximum temperature of 34 °C after 2 min. After that, the temperature starts to drop drastically to 30 °C and then continues increasing to 34 °C 2 min later. The same characteristic of the graph keeps repeating after 2 min later until it achieves 10 min total time taken of the graph. The total time taken for this analysis is 40 min.

4 Conclusion

In conclusion, the electrode and electrolytes for an aluminum-air battery have been selected based on the literature review and experimental results were obtained in the present study. It has been shown that the time of discharge is increasing gradually while the battery energy decreases inversely until the battery achieves the cut-off voltage that has been set. The higher alkali concentration helps to prevent the undesirable formation of aluminum hydroxide sediment and increases the power and energy capacity of the aluminum-air cell. A 1.5 V aluminum-air battery with a single cell and 4 cells with 3.88 V has been proven. Then followed by 4 cell temperature distribution analysis to see the temperature reading more drastically. Investigation at room temperature of 28 °C demonstrated that the temperature distribution of an aluminum-air battery with a 1 M concentration of NaOH as an electrolyte can rise obviously until to 34 °C maximum temperature.

The cheap cost estimation of the battery and characteristics of the aluminum-air battery developed in this study demonstrate the suitability of this battery as backup and emergency power sources, as well as power sources for electric vehicles. In addition, heat released during the operation of the aluminum-air battery can be

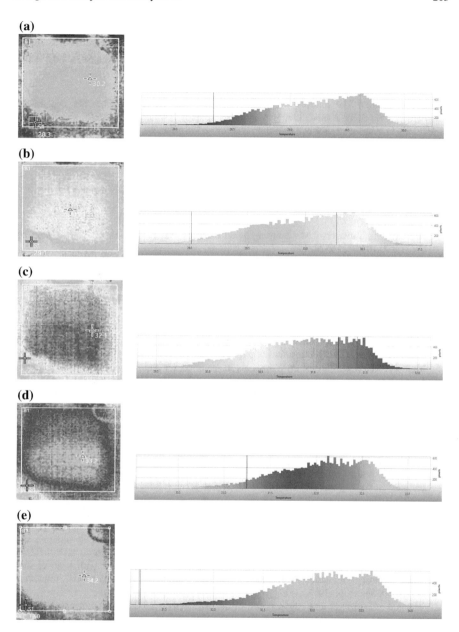

Fig. 9 **a** Thermography result for 0–10 s time taken. **b** Thermography result for 30 s time taken. **c** Thermography result for 50 s time taken. **d** Thermography result for 1 min 30 s time taken. **e** Thermography result for 2 min time taken

utilized for heating and climate control of the passenger compartment. Unlike traditional internal combustion engines, the aluminum-air battery generates no air pollutants and the product of anodic dissolution of aluminum can be incorporated and recycled in the standard.

Acknowledgments The authors would like to thanks to System Engineering and Energy Laboratory, University Kuala Lumpur Malaysian Spanish Institute and grateful thanks for the support given from supervisor and co-supervisor for giving guidance, ideas and endless motivation from the beginning until this project managed to complete.

References

1. Armand, M., Tarascon, J.M.: Building better batteries. Nature **451**, 652–657 (2008). https://doi.org/10.1038/451652a
2. Fan, L., Lu, H., Leng, J., Sun, Z., Chen, C.: The effect of crystal orientation on the aluminum anodes of the aluminum-air batteries in alkaline electrolytes. J. Power Sources **299**, 66–69 (2015). https://doi.org/10.1016/j.jpowsour.2015.08.095
3. Linse, C., Kuhn, R.: Design of high-voltage battery packs for electric vehicles. Advances in Battery Technologies for Electric Vehicles. Elsevier Ltd. (2015). https://doi.org/10.1016/b978-1-78242-377-5.00010-8
4. Liu, Y., et al.: A comprehensive review on recent progress in aluminum-air batteries. Green Energy Environ. (2017). https://doi.org/10.1016/j.gee.2017.06.006
5. Mokhtar, M., et al.: Recent developments in materials for aluminum-air batteries: a review. J. Ind. Eng. Chem. **32**, 1–20 (2015). https://doi.org/10.1016/j.jiec.2015.08.004
6. Mori, R.: Electrochemical properties of a rechargeable aluminum-air battery with a metal-organic framework as air cathode material. RSC Adv. **7**, 6389–6395 (2017). https://doi.org/10.1039/C6RA25164A
7. Nestoridi, M., et al.: The study of aluminum anodes for high power density Al/air batteries with brine electrolytes. J. Power Sources **178**, 445–455 (2008). https://doi.org/10.1016/j.jpowsour.2007.11.108
8. Pino, M., et al.: Carbon treated commercial aluminum alloys as anodes for aluminum-air batteries in sodium chloride electrolyte. J. Power Sources **326**, 296–302 (2016). https://doi.org/10.1016/j.jpowsour.2016.06.118
9. Rahman, M.A., et al.: High energy density metal-air batteries: a review. J. Electrochem. Soc. **160**, 1759–1771 (2013). https://doi.org/10.1149/2.062310jes
10. Wang, H.Z., et al.: A review on hydrogen production using aluminum and aluminum alloys. Renew. Sustain. Energy Rev. **13**, 845–853 (2009). https://doi.org/10.1016/j.rser.2008.02.009
11. Wang, J., et al.: Sulfur-mesoporous carbon composites in conjunction with a novel ionic liquid electrolyte for lithium rechargeable batteries. Carbon **46**, 229–235 (2008). https://doi.org/10.1016/j.carbon.2007.11.007
12. Wang, J.: Challenges and opportunities of nanostructured materials for aprotic rechargeable lithium-air batteries. Nano Energy **2**, 443–467 (2013). https://doi.org/10.1016/j.nanoen.2012.11.014
13. Zhang, X., et al.: Recent progress in rechargeable alkali metal-air batteries. Green Energy & Environ. **1**, 4–17 (2016). https://doi.org/10.1016/j.gee.2016.04.004

Corrosion Analysis of Aluminum-Air Battery Electrode Using Smoothed Particle Hydrodynamics

Faizah Osman, Amir Hafiz Mohd Nazri, Mohamad Sabri Mohamad Sidik and Muhamad Husaini Abu Bakar

Abstract Aluminum-air (Al-air) battery becomes one of the demanding batteries to power up an electronic device in our daily life. However, the corrosion behaviour of aluminium anodes is a major issue that must be carefully considered in the Al-air battery. This study is aimed to develop an Al-air battery single cell model and to simulate the corrosion by using the Smoothed Particle Hydrodynamics (SPH) method. The rate of corrosion at the anode and the effect of this corrosion to the performance of the Al-air battery is being study. As a result, the velocity profile of the anode corrosion and the electrolyte flow has been determined. These two measured parameters are closely significant toward corrosion behaviour. Thus, it has been proven that the SPH method is capable of modelling and simulating the corrosion behaviour in Al-air batteries.

Keywords Aluminum-air battery · Smoothed particle hydrodynamic (SPH) · Metal-air battery · Corrosion · Sediment movement

1 Introduction

In the new era of technology, batteries are under a huge demand to power up an electronic device in our daily life. The most common kind of battery used in modern passenger and freight battery electric vehicles (BEVs) are lithium-air

F. Osman · A. H. Mohd Nazri · M. S. Mohamad Sidik · M. H. Abu Bakar (✉)
System Engineering and Energy Laboratory, Universiti Kuala Lumpur Malaysian
Spanish Institute, Kulim Hi-Tech Park, 09000 Kulim, Kedah, Malaysia
e-mail: muhamadhusaini@unikl.edu.my

F. Osman
e-mail: faizah.osman23@s.unikl.edu.my

A. H. Mohd Nazri
e-mail: amir.nazri19@s.unikl.edu.my

M. S. Mohamad Sidik
e-mail: msabri@unikl.edu.my

© Springer Nature Switzerland AG 2019
M. H. Abu Bakar et al. (eds.), *Progress in Engineering Technology*,
Advanced Structured Materials 119, https://doi.org/10.1007/978-3-030-28505-0_18

217

batteries [1]. Every device has their own problem and same goes to lithium-air battery. Hence, Al-air batteries have been proposed as one of the alternative energy storage devices [2]. An Al-air battery is one of the metal air battery that theoretically provides more power than a regular battery and even can power up an electric vehicle [3]. The theoretical specific energy of an Al–air battery with an alkaline electrolyte can be as high as 200 Wh kg^{-1}, and with a neutral salt solution, it is between 300 and 500 Wh kg^{-1} [4]. However, there are unresolved problems of this battery that the anode is still facing a high rate of corrosion and negatively affects the potential of the battery [5].

The Al-air battery is mainly composed of an aluminum alloy electrode, an air electrode and an alkaline electrolyte. Its working principle is that the aluminum electrode (cathode) reacts with the hydroxyl ions of the electrolyte, generates Al(OH)$^{4-}$ and releases electrons. Electrons flow through the external line into the air electrode, which react with water and generate OH$^-$. The electrons in the external circuit are continuously oriented to form the current [2].

The specific electrode reaction and the battery reaction is as follows:

$$\text{Cathodic reaction:} \ 0.5O^2 + H_2O + 2e^- \rightarrow 2OH^- \tag{1}$$

$$\text{Anodic reaction:} \ Al \rightarrow Al_3 + + 3e^- \tag{2}$$

$$Al^3 + + 4OH^- \rightarrow Al(OH)^{4-} \tag{3}$$

A new generation of computational fluid dynamics (CFD) solver that has been developed is the Smoothed particle hydrodynamics (SPH) method for complex hydrodynamic problems that widely used in engineering problems for decades [6]. Meshless methods use to solve complicated domain topologies and multiphase systems in computational acoustics. To model fluid-structure interaction and the interaction between fluid particles and a solid, SPH is using the application of a fictitious repulsive force method to prevent the fluid particles from penetrating to solid form [7]. By using the SPH method we can simulate and observe the behaviour of the electrolyte motion inside the battery casing during the battery discharge process and we can see how the anode corrodes during the corrosion process which occurs during the discharge process.

2 Methodology

The methodology of this research consists of 3 main stages as shown in Fig. 1:

(i) Numerical Analysis

This research used numerical simulation to investigate the corrosion rate and the electrolyte flow inside the Al-air battery by using DualSPHysics software which is

Fig. 1 Stages of methodology

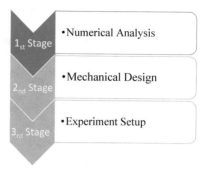

- 1st Stage — •Numerical Analysis
- 2nd Stage — •Mechanical Design
- 3rd Stage — •Experiment Setup

an open source CFD solver. The simulation results were compared with experiment results and calculation results.

(ii) Mechanical Design

Mechanical design consists of two developments of design. Firstly, the design of the Al-air battery by using the SolidWorks software because it offers complete 3D software tools that are easy to use and work together to help to design better products. Then, by using the Mastercam software, manufacturing processes is analyzed under one system to evaluate earlier designs earlier in the process to avoid unexpected costs and delays in finishing products on time. Secondly, several numerical models of the Al-air battery must be generated in the FreeCAD software for numerical analysis later.

(iii) Experiment Setup

In this section, the development of a simple experiment setup for the Al-air battery is required. The general purpose for the developing of the Al-air battery experiment setup is to measure the corrosion's distribution and corrosion's performance in the electrolytes. The electrolyte that been used in this experiment is sodium hydroxide (NaOH).

3 Results and Discussion

3.1 Weight Loss Method

After the experiment has been conducted for two hours, we can observe that the sediment is starting to form and flow with the flow of the reaction of aluminum corrosion as shown in Fig. 2. Besides the form of sediment, this reaction also released hydrogen gas as the product of the reaction. The weight of the aluminum plate was weighed, and a little weight loss occured from the experiment. This experiment has been conducted three times to observe the weight loss.

Fig. 2 Forming of sediment

Table 1 Weight loss with time

Time taken	Weight before (g)	Weight after (g)	Weight loss (g)
20 min	14.41	14.29	0.12
1 h	15.29	14.41	0.88
2 h	14.29	13.21	1.08

As we can see from the Table 1, the result of the experiment indicates that the plate lost 0.12 g after 20 min, 0.88 g after 1 h and 1.08 g after 2 h.

3.2 Simulation of Anode Corrosion

Figure 3 visualizes the motion of the battery during the reaction. The result of the simulation shows that the upper side of the sediment was deposited slowly. The simulation has run 500 sample of particle movement for five seconds of simulation. There are particle's movement of NaOH solution while the corrosion process occurs. The density of the fluid is set to 1000 kg/m^3 while the sediment density was set to 1500 kg/m^3 to observe the movement of the sediment. Other settings that have been set by the software for this simulation were not changed. The time taken to execute the simulation is depending on the size of the geometry and the number of the particles used in the simulation. DualSPHysics software can define the velocity of the particle movement and visualizing particle's movement in the battery.

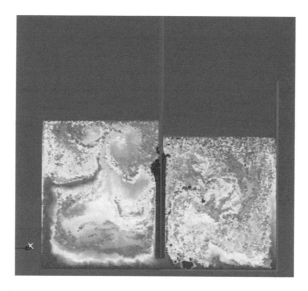

Fig. 3 Motion in the battery

3.3 *Velocity of Particles*

Figure 4 shows the velocity movement of the particle. The particle movement is tak-en from 50 particle movements at point 1294. The velocity value given are in form of X, Y and Z coordinates. From Fig. 4, we can observe that there is a positive and negative value due to the coordinate movement. It will show the velocity value from the last place the particle moves.

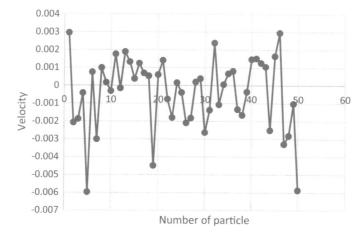

Fig. 4 Velocity of particles

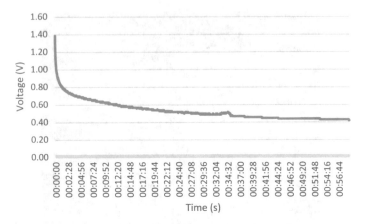

Fig. 5 Voltage over time

3.4 Battery Performance

The lifespan of a battery is very important to predict how long the battery can withstand the corrosion effect while producing energy. The graph in Fig. 5 shows that the voltage that has been produced by the battery is decreasing by the time. It is because of the corrosion effect that happened to the battery. The higher molarity of NaOH can make the aluminum anode corrode faster, and from the reaction, it produces high heat to the battery casing. It can be observed that the battery is facing a huge voltage drop at the first two minutes before dropping at a slower rate until the end of the testing. At minute 33.52 there is a sudden increase because of the

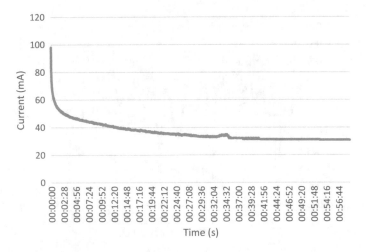

Fig. 6 Current over time

casing that has been used for the battery testing encountered a crack and leaked due to the high temperature that has been emitted during the reaction.

As the voltage drops, the current produced by the battery also drops at an equal rate with voltage. From the graph in Fig. 6, the current that has been produced also has dropped steadily during the discharge process. The battery achieved 98 milliamperes at the beginning and stopped at 30.86 milliamperes where the experiment has been stopped. The discharge rate is constant until the battery depleted and not producing energy anymore. The result shows that the current produced decreases direct proportional to the time.

4 Conclusions

The objective of this research was to model a single cell of an Al-air battery, to develop the SPH model for the battery and to evaluate the electrolyte behavior inside the battery. As we know that the SPH method will give us high computational time but, it will give us the accuracy of the simulation by consuming more time to develop the process. From the analysis, the behavior of the particles inside the battery is obtained. We can conclude that, when the corrosion process happen the electrolyte will flow toward the surface due to the bubbles that have been produced while the reaction process occurs, and the sediment will slowly sink into the bottom of the casing. The corrosion effect that is produced by the higher molar value of the electrolyte can affect the battery performance and life span due to the high reaction inside the battery.

Acknowledgements The authors gratefully acknowledge financial support for this work by the Universiti Kuala Lumpur Malaysian Spanish Institute (UniKL MSI) and System Engineering and Energy Laboratory (SEELab).

References

1. Tan, P., Kong, W., Shao, Z., Liu, M., Ni, M.: Advances in modeling and simulation of Li–air batteries. Prog. Energy Combust. Sci. (2017). https://doi.org/10.1016/j.pecs.2017.06.001
2. Cho, Y.-J., Park, I.-J., Lee, H.-J., Kim, J.-G. Aluminum anode for aluminum–air battery—Part I: influence of aluminum purity. J. Power Sources (2015). https://doi.org/10.1016/j.jpowsour.2014.12.026
3. Liu, Y., Sun, X., Sun, Q., Li, W., Li, J., Adair Keegan, R.: A comprehensive review on recent progress in aluminum-air batteries. Green Energy Environ. (2017). https://doi.org/10.1016/j.gee.2017.06.006
4. Mokhtar, M., Talib M.Z.M., Majlan, E.H., Tasirin, S.M., Ramli, W.M.F.W., Daud W.R.W., Sahari, J.: Recent developments in materials for aluminum-air batteries: a review. J. Ind. Eng. Chem. (2015). https://doi.org/10.1016/j.jiec.2015.08.004

5. Mutlu, R.N., Ateş, S., Yazıcı, B.: Al-6013-T6 and Al-7075-T7351 alloy anodes for aluminium-air battery. Int. J. Hydrogen. Energy (2017). https://doi.org/10.1016/j.ijhydene.2017.02.136
6. Shadloo, M.S., Oger, G., Le Touzé, D.: Smoothed particle hydrodynamics method for fluid flows, towards industrial applications : Motivations, current state, and challenges (2016). https://doi.org/10.1016/j.compfluid.2016.05.029
7. Zhang, A., Sun, P., Ming, F., Colagrossi, A.: Smoothed particle hydrodynamics and its applications in fluid-structure. J. Hydrodyn. (2017). https://doi.org/10.1016/s1001-6058(16)60730-8

Development of an Aluminum-Air Battery Using T6-6061 Anode as Electric Vehicle Power Source

Faizah Osman, Mohd Zulfadzli Harith, Mohamad Sabri Mohamad Sidik and Muhamad Husaini Abu Bakar

Abstract The demand for a long-lasting batteries for electric vehicle has increased throughout the years. The common rechargeable lithium-ion battery cannot fulfill this task. The Aluminum-air battery is an attractive candidate as it has a high power density. However, it has a high rate of corrosion on the anode electrode due to the aluminum reaction with the electrolyte. It becomes a challenge for researchers to make an ideal aluminum-air battery. The aim of this research is to determine the appropriate anode material's (pure aluminum or 6061-T6) for aluminum-air batteries. Two experiments, i.e. measuring the volume of hydrogen gas released and weight loss measurements of the aluminum metal before and after reaction with the electrolyte NaOH were performed using a specific experimental setup. The corrosion rate and the hydrogen gas evolution rate were calculated. Experiments were carried out with a molarity of 2, 3 and 4 M of NaOH with 10 min of immersion. The mass of specimens was measured by using an electronic digital weighing scale. As a result, it is found that the alloy elements enhance the corrosion resistivity of metal (aluminum).

Keywords T6-6061 · Aluminum-Air battery · Pure aluminum · Rate of corrosion · Hydrogen gas release

F. Osman · M. Z. Harith · M. S. Mohamad Sidik · M. H. Abu Bakar (✉)
System Engineering and Energy Laboratory, Universiti Kuala Lumpur Malaysian Spanish
Institute, Kulim Hi-Tech Park, 09000 Kulim, Kedah, Malaysia
e-mail: muhamadhusaini@unikl.edu.my

F. Osman
e-mail: faizah.osman23@s.unikl.edu.my

M. Z. Harith
e-mail: mzulfadzli.harith@s.unikl.edu.my

M. S. Mohamad Sidik
e-mail: msabri@unikl.edu.my

© Springer Nature Switzerland AG 2019
M. H. Abu Bakar et al. (eds.), *Progress in Engineering Technology*,
Advanced Structured Materials 119, https://doi.org/10.1007/978-3-030-28505-0_19

1 Introduction

Today's environment that we are living in is facing the extreme problem of pollution and global warming, and as a result, a lot of research in recent years has focused on the development of the environmentally friendly vehicle [1]. Air quality in developing cities has plunged and obviously is lower than the air quality in a rural area due to the emission of the fuel-powered vehicles. As modern society is encouraged to reduce the use of vehicles because of increasing pollution, a non-polluting vehicle such as an electric vehicle (EV) has become popular [2]. Even though the popularity of electric vehicles has risen, the power limit issue needs to be focused on if the number of users is to increase to a level at which a remarkable effect on environmental pollution can be achieved.

It was started in 1827, Anyos Jedlik, a Hungarian priest has built a simple viable electric motor with stator, rotor and commutator to power a small car. After eight years, a professor from the University of Groningen, the Netherlands has succeeded in building a small electric car [3]. Later on, a lot of technologies were introduced thanks to the studies on electric vehicle such as the primitive electric motor, early electric locomotive (with speed of only 6 km/h), the use of rail that conducts electric current and the introduction of the Lithium-ion battery as electric vehicle power source. In the latest studies, researcher have found a replacement for the common lithium-ion battery that is the aluminum-air battery. However, there is an unresolved problem of this battery that the anode is still facing a high rate of corrosion and this negatively affects the potential of the battery [4–6]. An Aluminum-air battery promises a lot of advantage for future automotive industry as it has a higher power density compared to the common battery used for our smartphone and laptop which is the lithium-ion battery with power density of 100–200 Wh/kg. On the other hand, the aluminum-air battery has about 400 Wh/kg [7, 8] meaning it has two to four times the power density of our common battery. However, there is still a lot of improvement needed to be done for it to be a good power source for EV.

This battery has a lot of components needed to be improved and one of it is the high corrosion rate of the anode in aqueous solution [4–6, 8]. The rate of corrosion is the speed at which any given metal deteriorates in a specific environment, as in this case is in acidic compound. The corrosion of the anode in an aluminum-air battery does decrease the efficiency of the battery as it causes the production of hydrogen gas and affects the battery negatively. Corrosion occurs at the oxidation reaction at the anode, where aluminum (Al) particles fuse with hydroxide (OH−) ions forming aluminum hydroxide (AlOH3) and dissolves, thus cannot operate as anode for the battery and needs to be replaced [9, 10].

Scientists and researchers have come out with a few methods to reduce the corrosion at the anode of this battery. Those methods are using alloyed aluminum instead of pure aluminum [5, 6], and by applying inhibitors [11] to the anode of the battery. The step of using alloys shows a great improvement compared to pure aluminum, but the corrosion problem still preventing the battery from achieving the full potential because of the extreme mass loss that's more than necessary for energy production. While an inhibitor is a great way to reduce the corrosion [5, 6] at the anode or to a metal as it acts as a protective layer on metal, as in this case it will be on the anode of the battery.

2 Methodology

In this study, the experimental setup was designed to carry out the experiment to evaluate the corrosion of metal (aluminum) and hydrogen released in NaOH solution. Two type of materials, i.e. pure aluminum and T6-6061 were compared. Designs were made by using the SolidWorks software and Mastercam V9. The anode electrode was fabricated to the actual product.

2.1 Hydrogen Gas Release Experiment

The main objective of this experiment is to evaluate the rate of hydrogen gas release and compare it with the rate of corrosion of each metal (T6-6061 and pure aluminum). This experiment was carried out with a variety molarity of NaOH in the range from 1 to 4 Mol of NaOH. The molarity of the NaOH was determined with a few fixed values: volume of water (0.1 L), molar mass of NaOH (40 g/mol). The molarity of NaOH was determined using the following equation.

$$M = \frac{n}{V} \tag{1}$$

where:

M molarity of mixture (M)
n number of mol of the compound (mol)
V volume of solvent (water) (L).

$$n = \frac{m}{\mathcal{M}} \tag{2}$$

where:

m mass of compound (NaOH) (g)
$\mathcal{M}$ molar mass of compound (NaOH) (g/mol).

The rate of hydrogen gas release was determined by using the following equation:

$$G = \frac{V}{AT} \tag{3}$$

where:

G rate of hydrogen gas release (ml cm^{-2} min^{-1})
V volume of hydrogen gas (ml)
A total surface area of metal (cm^2)
T time of immersion (min).

The experiment was initiated with filling a half beaker with water. Then a measuring cylinder was filled with water and covered with a piece of plastic. The water filled cylinder was turned upside down and put into the beaker half-filled with water until the lid of the cylinder immersed in the water. A delivery tube was placed in the measuring cylinder and the other end of the delivery tube was connected to the glass container cover. 100 ml of 1 M NaOH was poured into the glass container. Next, the specimen (T6-6061 aluminum) with approximate $20 \times 20 \times 2$ (mm) was put in the glass container and was quickly covered by the container cover. The specimen was immersed in NaOH for 10 min and the volume of water displaced in the measuring cylinder was recorded. The experiment was repeated three times and the average volume of water displaced was calculated to ensure the result. The experiment was repeated with 2, 3 and 4 M of NaOH, then using pure aluminium as a specimen for the experiment. The data was recorded, and the rate of hydrogen release was calculated.

2.2   Rate of Corrosion Experiment

The main objective of this experiment is to compare the rate of corrosion between the metals (T6-6061 and pure aluminum) and their rate of hydrogen gas release. The rate of corrosion was determined by using the following equation.

$$Corrosion\ rate = \frac{8.76\ W}{\rho AT} \tag{4}$$

where:

Corrosion rate   rate of corrosion of the metals (mg cm^{-2} h^{-1})
8.76   corrosion constant
W   weight difference of metal (mg)
ρ density of metal (g cm^{-3})
A total surface area of metal (cm^2)
T time of immersion (hour).

The experiment started with a specimen (T6-6061 aluminum) with a dimension approximately $20 \times 20 \times 2$ (mm) was weighed and recorded. Then, 100 ml of

1 M NaOH was poured into the glass container. The specimen was immersed in the NaOH in the glass container for 10 min. The specimen was taken out from the beaker and cleaned using tissue paper. The specimen was weighed again after 10 min of immersion inside NaOH and the weight of the specimen was recorded. The experiment was repeated three times and the mass difference was calculated to ensure a more accurate result. The experiment was repeated using NaOH with the molarity of 2, 3 and 4 M, then using pure aluminium as specimen. The data was recorded and the rate of corrosion was calculated.

3 Results and Discussion

3.1 Rate of Hydrogen Gas Release

Figure 1 shows the molarity of NaOH versus the rate of hydrogen gas release. Based on the graph, the data generally shows that the pure aluminum produces a greater volume of hydrogen gas compared to T6-6061 aluminum. Other than that, the results indicate that the rate of hydrogen gas release of pure aluminum is higher for every concentration of NaOH tested when compared to T6-6061 aluminum. Other than that, 1 M of NaOH shows a low rate of hydrogen gas released, but with increasing concentration of NaOH, the rate of hydrogen gas released also increased. This indicates that the higher concentration of NaOH, the greater the chemical reaction on T6-6061 aluminum and pure aluminum, thus speeds up the rate of hydrogen gas release.

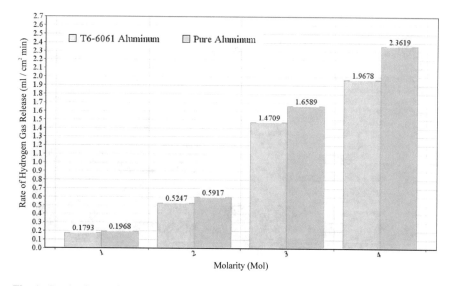

Fig. 1 Graph of rate of hydrogen release over molarity

This is because in lower concentration of NaOH, the NaOH to water ratio is lower, this means that the reaction of NaOH at a time is slow regarding the low presence of NaOH. The other way around when the concentration of NaOH is high, the NaOH to water ratio is higher allowing a high reaction at a time.

3.2 Rate of Corrosion

Figure 2 shows the concentration of NaOH versus rate of corrosion. Based on the graph, the data generally shows that the pure aluminum loses a greater amount of weight compared to T6-6061 aluminum. Other than that, the results indicate that the rate of corrosion of pure aluminum is higher for every concentration of NaOH tested when compared to T6-6061 aluminum even though the weight loss was the same at 3 M NaOH. Other than that, 1 M of NaOH shows a low rate of corrosion, but with increasing concentration of NaOH, the rate of hydrogen gas released also increased.

3.3 Overall

Based on both experiments, the result recorded came out with almost the same pattern. The result shows a lower rate at a lower concentration of NaOH and increased rate with increasing concentration of NaOH for both experiments.

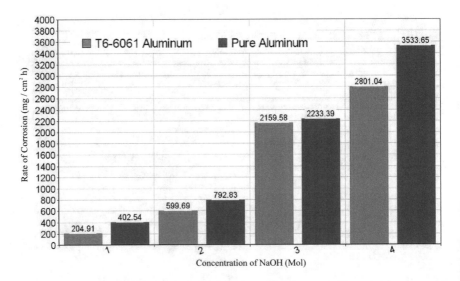

Fig. 2 Graph of rate of corrosion versus concentration of NaOH

Even the significant increase at 3 M of NaOH is similar in both experiments. It can be concluded that the volume of hydrogen gas released and the rate of hydrogen gas release indicates the rate of corrosion of the metal specimens. The higher the rate of hydrogen release, the higher the corrosion rate of the metal. The rate of corrosion and rate of hydrogen gas release might vary depending on the electrolyte used in the experiment.

4 Conclusions

The experiments show that the concentration of electrolyte affects mainly the rate of chemical reaction with metal, as in this case the rate of corrosion of aluminum. From the results that were discussed earlier, it can be concluded that T6-6061 aluminum has a better corrosion resistivity compared to pure aluminum. T6-6061 aluminum corrodes slower than pure aluminum, proofed by the lower rate of corrosion and lower rate of hydrogen gas release. This experiment indicates that the alloying element of aluminum improves the resistivity towards corrosion.

Aluminum has been studied by researchers for their high power density compared to the lithium in order to produce a battery that provides a longer continuous power supply. As pure aluminum has a higher rate of corrosion, it is less suitable to be applied as anode for the aluminum-air battery. Since the corrosion is high it will completely corrode the aluminum before it reaches the theoretically high power density. T6-6061 aluminum has a higher potential to be applied as aluminum-air battery anode because of the good corrosion resistance. This will result in an anode that tends to corrode slower and has a high potential to utilize its high power density that supports battery for a longer power supply time.

Acknowledgements The authors gratefully acknowledge financial support for this work by the Universiti Kuala Lumpur Malaysian Spanish Institute (UniKL MSI) and System Engineering and Energy Laboratory (SEELab).

References

1. Lonngren, K.E., Bai, E.-W.: On the global warming problem due to carbon dioxide. Energy Policy (2008). https://doi.org/10.1016/j.enpol.2007.12.019
2. Andwari, A.M., Pesiridis, A., Rajoo, S., Martinez-Botas, R., Esfahanian, V.: A review of battery electric vehicle technology and readiness levels. Renew. Sustain. Energy Rev. (2017). https://doi.org/10.1016/j.rser.2017.03.138
3. Garche, J., Moseley, P.T., Karden, E.: Lead–acid batteries for hybrid electric vehicles and battery electric vehicles. Adv. Batter. Technol. Electr. Veh. (2015)https://doi.org/10.1016/b978-1-78242-377-5.00005-4
4. Kumari, P.D.R, Jagannath, N., Shetty, A.N.: Corrosion behavior of 6061/Al-15 vol. pct. SiC (p) composite and the base alloy in sodium hydroxide solution. Arab. J. Chemistry. (2016). https://doi.org/10.1016/j.arabjc.2011.12.003

5. Huang, I.-W., Hurley, B.L., Yang, F., Buchheit, R.G.: Dependence on temperature, pH, and Cl − in the uniform corrosion of aluminum alloys 2024-T3, 6061-T6, and 7075-T6. Electrochim Acta (2016). https://doi.org/10.1016/j.electacta.2016.03.125
6. Yang, L., Wan, Y., Qin, Z., Xu, Q., Min, Y.: Fabrication and corrosion resistance of a graphene-tin oxide composite film on aluminium alloy 6061. Corros. Sci. (2018). https://doi.org/10.1016/j.corsci.2017.10.031
7. Mutlu, R.N., Ateş, S., Yazıcı, B.: Al-6013-T6 and Al-7075-T7351 alloy anodes for aluminium-air battery. Int. J. Hydrog. Energy (2017). https://doi.org/10.1016/j.ijhydene.2017.02.136
8. Liu, Y., Sun, X., Sun, Q, Li, W., Li, J., Adair, K.R.: A comprehensive review on recent progress in aluminum-air batteries. Green Energy & Environ. (2017). https://doi.org/10.1016/j.gee.2017.06.006
9. Yuan, B., Tan, S., Liu, J.: Dynamic hydrogen generation phenomenon of aluminum fed liquid phase Ga–In alloy inside NaOH electrolyte. Int. J. Hydrog. Energy (2016). https://doi.org/10.1016/j.ijhydene.2015.10.044
10. Zhang, X., Wang, X.-G., Xie, Z., Zhou, Z.: Recent progress in rechargeable alkali metal–air batteries. Green Energy Environ. (2016). https://doi.org/10.1016/j.gee.2016.04.004
11. Verma, C., Singh, P., Bahadur, I., Ebenso, E.E., Quraishi, M.A.: Electrochemical, thermodynamic, surface and theoretical investigation of 2-aminobenzene-1,3-dicarbonitriles as green corrosion inhibitor for aluminum in 0.5 M NaOH. J. Mol. Liquids. (2015). https://doi.org/10.1016/j.molliq.2015.06.039

Synthesis and Thermal Characterization of Graphite Polymer Composites for Aluminium Ion Batteries

Faizatul Azwa Zamri, Najmuddin Isa, Muhamad Husaini Abu Bakar and Mohd Nurhidayat Zahelem

Abstract Phenolic resin is a thermosetting polymer resin is that is known for its excellent thermal properties and chemical stability. Thus, it could be advantageous if it could be utilized in graphite polymer composites as the cathode in aluminium ion batteries. In this study, graphite polymer composites with 3 cm of diameter and 3 cm of thickness had been fabricated and then their thermal characteristics were determined. The composition ratios of 40/60, 50/50, 60/40 and 70/30 by their weight percentage (wt%) of graphite/phenol respectively were fabricated using the hot compression method. Any further increment of graphite percentage in the composition would produce a very fragile composite. In the physical properties characterization, the bulk density, true density, and porosity percentage were determined. Besides, the thermal conductivity of the graphite composites had been measured according to ASTM E1350 standard for thermal characterization. It is observed that an increase in graphite content results in an increase in porosity content and thus reducing its thermal conductivity. It is concluded that a composition of 40/60 and 50/50 by its %wt of graphite/phenol has good physical and thermal properties with ~ 1.7 g/cm^3, 6–8%, and ~ 36 W/mK of bulk density, porosity percentage and thermal conductivity, respectively.

Keywords Graphite polymer composite · Graphite · Phenol · Thermal conductivity

F. A. Zamri · N. Isa · M. H. Abu Bakar (✉) · M. N. Zahelem
System Engineering and Energy Laboratory, Universiti Kuala Lumpur Malaysian Spanish Institute, Kulim Hi-Tech Park, 09000 Kulim, Kedah, Malaysia
e-mail: muhamadhusaini@unikl.edu.my

F. A. Zamri
e-mail: faizatul.zamri08@s.unikl.edu.my

N. Isa
e-mail: najmuddin031094@gmail.com

M. N. Zahelem
e-mail: mnurhidayat@unikl.edu.my

© Springer Nature Switzerland AG 2019
M. H. Abu Bakar et al. (eds.), *Progress in Engineering Technology*,
Advanced Structured Materials 119, https://doi.org/10.1007/978-3-030-28505-0_20

1 Introduction

Lithium-ion batteries have dominated the battery market of energy storage for portable electronics and smart grids and so on for two decades. Unfortunately, the commercial lithium-ion batteries (LIBs) so far cannot satisfy large scale storage applications due to the safety, cost and reliability issues. Consequently, great efforts have been devoted to explore new battery systems to satisfy the urgent need for sustainable and efficient energy storage in modern society [1]. Regarding this issue, new types of rechargeable battery systems were developed which could fuel broad applications from personal electronics to grid storage. Since then, aluminum-ion batteries have attracted much attention due to their low cost, environmental benignity, and high charge density [2].

Natural graphite is widely used as an electrode material, which has been commercialized successfully in the lithium-ion battery due to its unique physical and chemical properties [3]. In contrast, a graphite cathode was just recently successful developed for aluminum-ion batteries with comparable electrochemical properties as lithium-ion batteries [4]. However, achieving desirable electrode properties is still connected to many challenges [5]. Therefore, enhancement on the graphite materials as cathode still requires further development.

A composite material is known as two or more materials with different properties that are combined together. The constituents are combined in such a way that they keep their individual physical phases and are non-solvable in each other or do not form a new chemical compound [6]. It is known that polymer/graphite composites exhibit a high thermal conductivity and an electrical conductivity at a fairly low concentration [7]. Thus, it is expected that graphite/polymer composites will potentially be used as the cathode for aluminium-ion batteries.

Phenolic resins have an excellent affinity for graphite and other forms of carbon. Manufacturers often use the resin simply as a binder and adhesive for their carbon materials. At high temperature, phenolic resins form a char of amorphous carbon. This means that phenolic bonded carbon materials can be heat treated to yield an all carbon structure. Because of these unique properties, phenolic resins find application in the manufacture of electrodes, carbon composites, carbon seals, and washers [6]. However, the fabrication and characterization of the graphite/phenol is still unclear. Therefore, the synthesis of graphite/phenol composites with different composition ratio is developed in this research. Then, physical and thermal characterization will be performed in order to determine their properties.

2 Methodology

(i) Synthesis of graphite polymer composites

The graphite/phenol composites were prepared according to the composition ratios 40/60, 50/50, 60/40, 70/30 of graphite/phenol by its weight percentage with a total mass 39 g for each sample. Figure 1 shows the synthesis method of the graphite polymer composites. The samples were fabricated using the compression molding techniques using an AMP-5 automatic mounting press machine. A mold release agent (200-10006 PTFE Spray) was applied prior to the mounting cycle to avoid sticking on the mold surface. Figure 2 shows the fabricated samples.

(ii) Physical Characterization

The mass and dimension of the samples were measured to determine their bulk density according to Eq. (1). Then, the density test has been done to measure the true density of the samples. The porosity percentages were calculated according to Eq. (2).

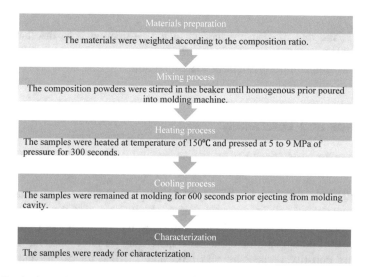

Materials preparation
The materials were weighted according to the composition ratio.

Mixing process
The composition powders were stirred in the beaker until homogenous prior poured into molding machine.

Heating process
The samples were heated at temperature of 150°C and pressed at 5 to 9 MPa of pressure for 300 seconds.

Cooling process
The samples were remained at molding for 600 seconds prior ejecting from molding cavity.

Characterization
The samples were ready for characterization.

Fig. 1 Synthesis method of graphite/phenol composites

Fig. 2 Graphite/phenol composites

$$Bulkdensity = \frac{mass(\text{g})}{volume(\text{cm}^3)} \qquad (1)$$

$$Porosity(\%) = \frac{Bulkdensity - Truedensity}{Truedensity} \times 100\% \qquad (2)$$

(iii) Thermal Characterization

In this section, the thermal conductivity test was performed according to the ASTM E1350 standard to determine their thermal characterization. The thermal conductivity, k, of graphite polymer composites was calculated according to the Fourier's law for linear conduction (see Eq. 3),

$$Q = -kA(dT/dx) \qquad (3)$$

where Q is the rate of heat conduction in the x-direction, k is the thermal conductivity of the material, A is the cross-sectional area normal to the x-direction, and dT/dx is the temperature gradient in the x-direction.

3   Results and Discussion

3.1   Density and Porosity

Table 1 shows the density and porosity of the graphite-polymer composite based on the composition ratio. It is observed that the bulk density of the graphite/phenol composite obtained was about 1.7 g/cm^3. Besides, the percentage of porosity in the graphite/phenol composite increased with increasing graphite content. The lowest and highest porosity percentage was 6.63 and 13.70% for 50/50 and 70/30 of graphite/phenol composition, respectively.

Table 1 Density and porosity of graphite/phenol composites

Graphite/phenol composition	Bulk density (g/cm^3)	True density (g/cm^3)	Porosity (%)
40/60	1.70	1.58	7.61
50/50	1.67	1.57	6.63
60/40	1.73	1.59	8.77
70/30	1.73	1.52	13.70

Table 2 Density and porosity of graphite/phenol composites

Power, Q (W)	Thermal conductivity, k (W/mK)			
	40/60	50/50	60/40	70/30
5	43.18	41.64	27.57	26.24
10	40.61	40.10	27.58	26.85
15	36.12	36.53	25.53	24.40
20	32.96	34.11	22.83	22.54
Average	38.22	38.09	25.88	25.01

3.2 Thermal Conductivity

Table 2 shows the thermal conductivity at different power. The thermal conductivity value was calculated by averaging the value from the test using 5, 10, 15, and 20 watts. The highest thermal conductivity was 38.22 W/mK obtained for 40% of graphite content in the graphite-polymer composite. However, the thermal conductivity of the composites was decreased with increasing of graphite content in the composition.

4 Conclusions

The objective of this research was to fabricate graphite/phenol composites with a different composition. Then, the physical properties including bulk density, true density and porosity were measured. Lastly, the thermal conductivity test was performed to calculate their thermal conductivity value, k. The composite with 40/60, 50/50, 60/40 and 70/30 by its wt% was successfully fabricated by the hot compression method. However, any further increment of graphite content would produce a very fragile composite. It is concluded that the increasing graphite content in the graphite/phenol composites results in an increased porosity which reduced their thermal conductivity. Besides, the composition of 40/60 and 50/50 by its %wt of graphite/phenol has good physical and thermal properties with ~ 1.7 g/cm^3, 6–8%, ~ 36 W/mK of bulk density, porosity percentage and thermal conductivity, respectively.

Acknowledgements The authors gratefully acknowledge financial support for this work by the Universiti Kuala Lumpur Malaysian Spanish Institute (UniKL MSI) and System Engineering and Energy Laboratory (SEELab).

References

1. Wang, S, Jiao, S., Song, W.L., Chen, H.S., Tu, J., Tian, D., Jiao, H., Fu, C., Fang, D.N.: A novel dual-graphite aluminum-ion battery. Energy Storage Mater. (2018). https://doi.org/10.1016/j.ensm.2017.12.010

2. Zhang, E., Cao, W., Wang, B., Yu, X., Wang, L., Xu, Z., Lu, B.: A novel aluminum dual-ion battery. Energy Storage Mater. **11**, 91–99 (2018). https://doi.org/10.1016/j.ensm.2017.10.001
3. Wei, J., Chen, W., Chen, D., Yang, K.: An amorphous carbon-graphite composite cathode for long cycle life rechargeable aluminum ion batteries (2017). https://doi.org/10.1016/j.jmst.2017.06.012
4. Lin, M.C., Gong, M., Lu, B., Wu, Y., Wang, D.Y., Guan, M., Angell, M.: An ultrafast rechargeable aluminium-ion battery. Nature **520**, 324–328 (2015). Macmillan Publ. Limited. https://www.nature.com/articles/nature14340
5. Bhauriyal, P., Mahata, A., Pathak, B.: The staging mechanism of AlCl4 intercalation in a graphite electrode for an aluminium-ion battery. Phys. Chem. Chem. Phys. **19**, 7980–7989 (2017). http://dx.doi.org/10.1039/C7CP00453B
6. Hossain, J., Saiful Alam, M., Chandra Paul, S., Islam, S.: Fabrication and characterization of carbon black and phenol formaldehyde composites. Ind. Chem. **04**(01) (2018). http://doi.org/10.4172/2469-9764.1000127
7. Mokhena, T.C., Mochane, M.J., Sefadi, J.S., Motloung, S.V., Andala, D.M.: Thermal conductivity of graphite-based polymer composites. Impact of Thermal Conductivity on Energy Technologies (2018). http://dx.doi.org/10.5772/intechopen.75676

Design and Analysis of an Aluminium Ion Battery for Electric Vehicles

Faizatul Azwa Zamri, Mohamad Zhairul Faris Jumari, Muhamad Husaini Abu Bakar and Mohd Nurhidayat Zahelem

Abstract The Lithium-ion battery has been widely used in the development of electric vehicles. However, awareness of battery safety has been encouraged to use aluminium based batteries. Therefore, in this study, an aluminium ion battery cell with 25 mm × 100 mm of diameter and height respectively was designed using the SolidWorks 2016 software. Then the aluminium-ion battery was fabricated using different electrolyte types including potassium hydroxide (KOH), sodium hydroxide (NaOH) and a mixture of sodium hypochlorite with sodium hydroxide (NaOCl + NaOH) to determine the battery characteristics. The battery characteristics were obtained using an Arduino battery performance tester connected to the PLX-DAQ software as an interface. A thermography test was also performed to observe the heat distribution on the outer surface of the battery using a thermal imager model U5855A TruIR thermal imager connected to the TrueIR Analysis and Reporting Tool software. It is observed that aluminium-ion battery using a mixture of sodium hypochlorite with sodium hydroxide provides good battery characteristics which specific voltage and current density obtained as 1.13 V for and 79.31 mA respectively with 8 h operating time. The heat distribute over the surface was moderate with the highest temperature of 37 °C and constantly for 15 min. As a conclusion, an aluminium-ion battery has been developed for future review in electric vehicles application.

Keywords Aluminum-ion batteries · Specific voltage · Specific current density · Heat distribution

F. A. Zamri · M. Z. Faris Jumari · M. H. Abu Bakar (✉) · M. N. Zahelem
System Engineering and Energy Laboratory, Universiti Kuala Lumpur Malaysian Spanish Institute, Kulim Hi-Tech Park, 09000 Kulim, Kedah, Malaysia
e-mail: muhamadhusaini@unikl.edu.my

F. A. Zamri
e-mail: faizatul.zamri08@s.unikl.edu.my

M. Z. Faris Jumari
e-mail: fzhairul@gmail.com

M. N. Zahelem
e-mail: mnurhidayat@unikl.edu.my

© Springer Nature Switzerland AG 2019
M. H. Abu Bakar et al. (eds.), *Progress in Engineering Technology*,
Advanced Structured Materials 119, https://doi.org/10.1007/978-3-030-28505-0_21

1 Introduction

Aluminum-ion battery is the new technology that is used to replace the lithium-ion battery in the future since it has high energy stored on a per volume basis because an ability to exchange three electrons during the electrochemical process as shown in Eq. (1). Hence, the design of the battery is expected to be in a smaller size [1]. Aluminum, in fact, possesses the highest volumetric capacity, 8040 mAh cm^{-3}, which is four times higher than lithium and has a good gravimetric capacity of 2980 mAh [2]. Additional, aluminium is the most abundant metal element in the earth's crust expected to produce low-cost batteries for large scale applications including electric vehicles [1, 2].

$$Al^{3+} + 3e^- \leftrightarrow Al \tag{1}$$

Lithium-ion batteries gradually increased their significant important as power sources for electric vehicles due to high theoretical capacity and good cycle-life [3]. However, the lithium-ion battery has an issue which is to ensure that lithium-ion particles can move easily between electrodes, volatile and flammable chemical compounds that are pressurized inside the battery cells [4, 5]. Since aluminium can be handled open air leading to enormous advantages for cell fabrication, thus it extremely improves the safety level of electrochemical storage systems [2]. Therefore, an aluminium-ion battery is developed and its battery characteristics were evaluated.

To date, aluminium, graphite and AlCl3/1-ethyl-3-methylimidazolium chloride [EMIm]Cl ionic liquid as anode, cathode, and electrolytes respectively have reported to afford unprecedented discharge voltage profiles, cycling stabilities and rate capabilities in the rechargeable aluminium-ion battery development, [6, 7, 9–12]. However, the cost and preparation for the ionic liquid are high. Therefore, it is advantageous to find alternative electrolytes with low cost and require simple preparation. The battery thermal management is critical in achieving performance and extended life of batteries in electric and hybrid vehicles under real driving conditions. Appropriate modeling for predicting the thermal behavior of battery systems in vehicles helps to make decisions for improved design and shortens the development process [8]. Therefore, it is important to understand the thermal distribution on the outer surface of an aluminium-ion cell to ensure battery safety for future battery development electric vehicles applications.

2 Methodology

(i) Battery development:

The first stage of this work is battery development (Fig. 1). The battery was designed based on the motivation to produce a small and compact size battery.

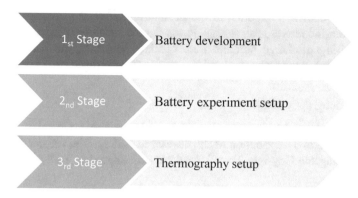

Fig. 1 Stages of methodology

Thus, the several designs were modeled using the SolidWorks 2016 software. The graphite rod with 10 mm of its diameter and 18 mm of aluminum rod were used as the cathode and anode of the battery respectively. A 1 M concentration of three different electrolytes, including potassium hydroxide (KOH), sodium hydroxide (NaOH) and the mixture of sodium hypochlorite with Sodium Hydroxide (NaOCl + NaOH), was used for testing.

(ii) Battery experiment setup

In this section, an Arduino Nano battery performance tester (Fig. 2) with 10 W load applied was used to study the characteristic of the battery during the discharge

Fig. 2 Battery performance tester kit

process. The data was exported to the PLX-DAQ software to plot the discharge curves. From the curves, the voltage across the load resistor divided by the resistance resulting discharge currents were obtained and then the battery capacity (milliamp-hour) value was calculated by multiplying the discharge current with the time taken.

(iii) Thermography setup

The thermal imaging test had been done during the discharging process to observe the thermal distribution of the developed battery. The recorded video was analyzed using the Keysight TrueIR Analysis and Reporting Tool software.

3 Results and Discussion

3.1 Design of the Battery

Several design concepts were developed as shown in Fig. 3. The battery design (a) has been selected for fabrication since it meet all criteria such as simplicity, high surface area, low cost, and easy to manufacture. It is a cylindrical shape with 10 mm × 100 mm of its diameter and height respectively. The other shell is made from aluminium (anode) and graphite rod (cathode) and electrolyte were assembled inside.

3.2 Battery Discharging Curves

Figure 4 shows the discharging curves of aluminium-ion battery with different electrolytes and Table 1 summarizes their capacities. It is observed that the mixture of sodium hypochlorite and sodium hydroxide had the highest battery capacity with 8 h operating time as shown in Table 1. However, potassium hydroxide has longer operating time with the lowest battery capacity. The specific capacity and specific current density are similar for all types of electrodes which was 65 mAh/g and 560 mA/g, respectively.

3.3 Thermal Distribution Analysis

Figure 5 shows the graph of the thermal distribution against time for an aluminium-ion battery using a mixture of sodium hypochlorite and sodium hydroxide as the electrolyte. The highest temperature observed on the outer surface

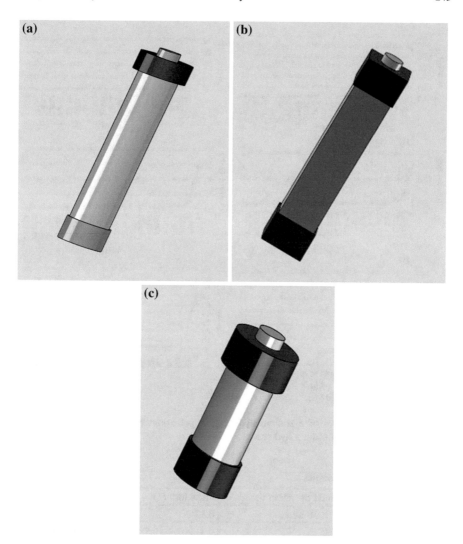

Fig. 3 Several battery design **a** cylindrical battery with 100 mm of its length **b** square battery and **c** cylindrical shape battery with 50 mm of its length

of the battery was 37 °C which is relatively moderate compared to ambient temperature which was 30 °C. This is requires more design optimization for better heat dissipation.

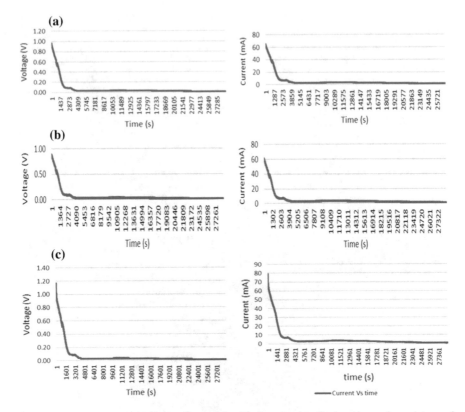

Fig. 4 Discharging curves of **a** sodium hydroxide **b** potassium hydroxide, and **c** mixture of sodium hypochlorite with sodium hydroxide

Table 1 Battery characteristics

Electrolyte	Total operation time	Mass loss (g)	Capacity
NaOH	7 h 30 min	0.13	0.96 V at 65.20 mA
KOH	9 h	0.1	0.89 V at 60.97 mA
NaOCl + NaOH	8 h	0.1	1.13 V at 79.31 mA

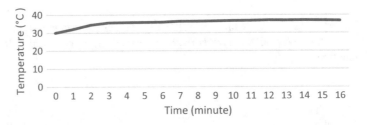

Fig. 5 Thermal distribution

4 Conclusions

The objective of this research is to model the single cell of an Al-ion battery, evaluate battery discharging characteristics using potassium hydroxide (KOH), sodium hydroxide (NaOH) and a mixture of sodium hypochlorite with sodium hydroxide (NaOCl + NaOH). Lastly, the thermal distribution was observed on the surfaces of the battery. It is known that the specific capacity and current density of the developed battery was 65 mAh/g and 560 mA/g, respectively. The mixture of NaOCl + NaOH electrolyte provides good characteristics which were 1.13 V for and 79.31 mA for voltage and current density respectively for 8 h of operating time. Besides, the thermography test shows that the temperature was rising from ambient temperature of 30 to 37 °C and remained for 15 min. This shows that the heat dissipated moderately to the outer surface. Lastly, this review has summarized recent developments of Al anode, graphite cathode, and electrolytes of an aluminum-ion battery which is expected to be useful in the electric vehicles battery development.

Acknowledgements The authors gratefully acknowledge financial support for this work by the Universiti Kuala Lumpur Malaysian Spanish Institute (UniKL MSI) and System Engineering and Energy Laboratory (SEELab).

References

1. Das, S.K., Mahapatra, S., Lahan, H.: Aluminium-ion batteries: developments and challenges. J. Mater. Chem. A **5**(14), 6347–6367 (2017). https://doi.org/10.1039/C7TA00228A
2. Elia, G.A., Marquardt, K., Hoeppner, K., Fantini, S., Lin, R., Knipping, E., Hahn, R., et al.: An Overview and future perspectives of aluminum batteries. Adv. Mater. (2016). http://doi.org/10.1002/adma.201601357
3. Chen, Y., Evans, J.W.: Thermal analysis of lithium-ion batteries. J. Electrochem. Soc. **143**(9), 2708–2712 (1996). https://doi.org/10.1016/j.jpowsour.2004.05.064
4. Bandhauer, T.M., Garimella, S.: Passive, internal thermal management system for batteries using microscale liquid e vapor phase change. Appl. Therm. Eng. **61**(2), 756–769 (2013). https://doi.org/10.1016/j.applthermaleng.2013.08.004
5. Bandhauer, T.M., Garimella, S., Fuller, T.F.: A critical review of thermal issues in lithium-ion batteries. J. Electrochem. Soc. **158**(3) (2011). http://doi.org/10.1149/1.3515880
6. Lin, M.C., Gong, M., Lu, B., Wu, Y., Wang, D.Y., Guan, M., Angell, M., Chem, C., Yang, J., Joe-Hwang, B., Dai, H.: An ultrafast rechargeable aluminium-ion battery. Nature (2015). http://doi.org/10.1038/nature14340
7. Wang, D.Y., Wei, C.Y., Lin, M.C., Pan, C.J., Chou, H.L., Chen, H.A., Gong, M., Wu, Y., Yuan, C., Angell, M., Hsieh, Y.J., Cen, Y.H., Wen, C.Y., Chen, C.W., Hwang, B.J., Chen, C. C., Dai, H.: Advanced rechargeable aluminium ion battery with a high-quality natural graphite cathode. Nat. Commun. **8**, 1–7 (2017). https://doi.org/10.1038/ncomms14283
8. Pesaran, A., Kim, G.H.: Battery Thermal Management System Design Modeling. United States. https://www.osti.gov/biblio/894990-battery-thermal-management-system-design-modeling
9. Elia, G.A., Hasa, I., Greco, G., Diemant, T., Marquardt, K., Hoeppner, K., Urgen Behm, R.J., Hoell, A., Passerini, S., Hahn, R.: Insights into the reversibility of aluminum graphite batteries. J. Mater. A. (2017). http://dx.doi.org/10.1039/C7TA01018D

10. Wang, C., Li, J., Jiao, H., Tu, J., Jiao, S.: The electrochemical behaviour of an aluminium alloy anode for rechargeable Al ion batteries using an AlCl₃ urea liquid electrolyte. RSC Adv. **7**, 32288–32293 (2017). http://dx.doi.org/10.1039/C7RA05860H
11. Ng, K.L., Malik, M., Azimia, G.: A new generation of rechargeable aluminum ion battery technology. Advancing Solid State Electrochem. Sci. Technol. **85**(13), 199–206 (2018). http://dx.doi.org/10.1149/08513.0199
12. Wang, S., Jiao. S., Song. W.L., Chen, H.S., Tu J., Tian, D., Jiao, H., Fu, C., Fang, D.N.: A novel dual-graphite aluminum-ion battery. Energy Storage Mater. **12**, 119–127 (2018)

Automotive Metallic Component Inspection System Using Square Pulse Thermography

Nor Liyana Maskuri, Elvi Silver Beli, Ahmad Kamal Ismail
and Muhamad Husaini Abu Bakar

Abstract This paper presents an alternative inspection system for automotive components replacing the conventional way by using naked human eye and ultrasonic which are normally quiet time consuming. In order to solve the time consumption issue, we propose the thermography technique which has the capability to detect the internal defect in an efficient way and is proven to be one of the active thermography types. The process involves: (i) developing the square pulse thermography inspection system for automotive components. The time duration of heating is 7 min 10 s with 3002 sequence of an image, (ii) analyzing the effect of defect magnitude on the surface temperature distribution, and finally, (iii) determining a defect profile in a metallic element. The diameter of the defect is evaluated by calculating the ratio between the physical size and the pixel number. As a result, the defect of 12 artificial holes can be detected with less than 10% error. As a consequence, the proposed thermography method has a good potential to be utilized in an automotive inspection system.

Keywords Square pulse thermography network · Automotive component network · Discrete-Fourier transform network · Internal defect network · Temperature distribution network · Ultrasonic

N. L. Maskuri · E. S. Beli · A. K. Ismail · M. H. Abu Bakar (✉)
System Engineering and Energy Laboratory, University Kuala Lumpur-Malaysian Spanish
Institute, Kulim Hi-Tech Park, 09000 Kulim, Kedah, Malaysia
e-mail: muhamadhusaini@unikl.edu.my

N. L. Maskuri
e-mail: nliyana.maskuri@s.unikl.edu.my

E. S. Beli
e-mail: elvisilversilver@gmail.com

A. K. Ismail
e-mail: ahmadkamal@unikl.edu.my

© Springer Nature Switzerland AG 2019
M. H. Abu Bakar et al. (eds.), *Progress in Engineering Technology*,
Advanced Structured Materials 119, https://doi.org/10.1007/978-3-030-28505-0_22

1 Introduction

Quality monitoring is important when it comes to car manufacturer because a car comes with complex parts and each part needs to be inspected before it reaches to the assembly line. A defect will be the first parameter that need to be examined on the part. If the part is not inspected properly, there will be a lot time to waste to disassembly back the car to remove the part that has a defect. A car manufacturer normally uses external inspection to examine the car's part. This can be done using the raw human eye, and other non-destructive technique such magnetic particle testing and liquid penetrant testing but this techniques do not represent the internal defect of the part inspected.

Recently, the car manufacturers started are start implementing ultrasonic approaches, i.e. one of the non-destructive technique. However, the problem that a lot of manufacturing companies dealing with ultrasonic is this method required direct contact between transducer and the surface of the part that need to be inspected and it consumes a lot of time. When the inspection time comprises a huge part of the manufacture process, the productivity of the company will stagnant. The new method thermography technique was introduced to overcome this problem. This technique uses a infrared camera to evaluate the heat distribution from the car' component. As it is using infrared radiation, it will not require for direct contact to the part in order to inspect. Thermography also can inspect a wide surface in one single shot.

Automotive manufacturers use external inspection systems as a way to evaluate automotive components. These types of external inspection systems include using the raw human eyes and non-destructive techniques such as magnetic particle and liquid penetrant particle method. Raw human eyes are common and the most basic technique used in automotive industry because it both can save time and money. However, this technique is not representing the internal defect of the component. Internal defects are crucial because they can lead to unpredictable failure of the final product. Internal defects are usually caused by machine failure or human error.

Recently, automotive manufacturers have implemented ultrasonic techniques as an inspection approach for automotive part. The ultrasonic technique uses high frequency sound energy produced by the transducer to conduct the examination and make measurements. The sound energy is introduced and propagates through the materials in the form of waves. When there is a discontinuity (such as a crack) in the wave path, part of the energy will be reflected back from the flaw surface. The reflected wave signal is transformed into an electrical signal by the transducer and it is displayed on a screen. However, in order to inspect the component, the transducer must be in direct contact with the component and as a consequent the inspection process will consume a lot of time.

In order to overcome the problem, several objectives have been proposed (i) developing the square pulse thermography inspection system for automotive components. The time duration of heating is 7 min 10 s with 3002 sequence of an image, (ii) analyzing the effect of defect magnitude on the surface temperature distribution, and finally, (iii) determining a defect profile in a metallic element by using square pulse thermography.

This research is about to develop an automotive metallic component inspection system using the square pulse thermography. An aluminum alloy T6061 plate is used in this experiment. This material is used to represent a part of car's boot. The result of the analysis represents in form of graphs and images. The graph shows the correlation between the temperature and the pixel of the image, the magnitude of the defect should be appearing in this graph. In form of the image, the shape and type of defect should able to be identified. After that, the diameter of the defect can be estimated.

2 Literature Review

2.1 Non-destructive Test

Non-destructive testing refers to the evaluation and inspection process of materials or components for finding defects and flaws in comparison with some standard without altering the original attributes or harming the object being tested [1]. There are a lot of advantages of non-destructive testing when dealing with finding defects and it gives a lot of benefits especially for the manufacturing industry that needs to check their product quality. Non-destructive techniques make available or provide a cost effective means of testing of a sample for individual investigation and examination or may be applied on the whole material for checking in a production quality control system [2]. There are five types of non-destructive testing techniques that are usually used to inspect defects, magnetic particle testing, liquid penetrant testing, gamma-ray radiography and thermography.

2.2 Thermography

The principle of thermography is based on the physical phenomenon that objects with a temperature above the absolute zero (0 K or -273.15 °C), emit electromagnetic radiation depending on their temperature [3]. Infrared thermography is used to measure, or map surface temperatures based on the infrared radiation given off by an object as heat flows through, to or from that object. The majority of infrared radiation is longer in wavelength than the visible light but can be detected using thermal imaging devices, commonly called "infrared cameras." For accurate IR testing, the part(s) being investigated should be in direct line of sight with the camera, i.e., should not be done with panel covers closed as the covers will diffuse the heat and can result in false readings. Used properly, thermal imaging can be used to detect corrosion damage, delamination, disbonds, voids, inclusions as well as many other detrimental conditions.

2.3 Active Thermography and Passive Thermography

The two basic configurations for implementing an infrared thermographic technique, first is active and second is passive. In passive thermography, the monitoring of the thermal radiation, emitted by the surface of the test body under natural conditions, is used and it is widely applied as a standard quality control technique of historic structures since many years ago [4]. In the active approach, an external stimulus is required to generate relevant temperature differences not present otherwise. Known characteristics of this external stimulus (example: time t_0 when it is applied) enable quantitative characterization such as for instance the depth of a detected disbond. Depending on the external stimulus, different approaches of active thermography have been developed, such as pulse thermography (PT)) Fig. 1a, vibrothermography (VT) and Fig. 1b, lock-in thermography (LT) [5].

2.4 Image Processing

The raw data that we get from this experiment is in the form of a thermogram (jpg) in which the zone of the defect should appear. However, the defect zone sometimes appears with a subtle signature due to all factors that degrade infrared (IR) images from self-emission of the IR camera to the non-uniform properties of the surface where data are collected [6]. We need to perform the quantitative analysis for detection and characterization because of the signal in the thermal bands are intrinsically weak because liberated photonic energy due to the oscillatory nature of the particle inside the matter is inversely proportional to the wavelength. It is necessary to fix some problems related to the acquisition system.

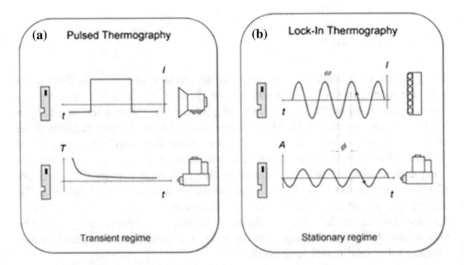

Fig. 1 a Pulsed thermography, **b** lock-in thermography

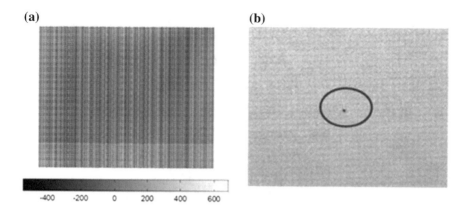

Fig. 2 **a** Fixed pattern noise, **b** bad pixel

The image tends to spoil caused by a few factors; the first factor is vignetting due to limited aperture. Vignetting is noise on thermograms that causes a darkening of the image corners with respect to the image center due to limited exposure [7]. Vignetting depends on both pixel location and temperature difference with respect to the ambient and this make it more complex effect to address since it is related to the temperature of the scene with respect to temperature of the lens with no effect present when both are identical. The second factor is the fixed pattern noise (FPN) which is the result of a difference in responsivity of the detector to incoming irradiance. It is a common problem when working with focal plane arrays (FPA). FPN for a particular configuration can be recovered from a blackbody image for later subtraction from the thermogram sequence. Figure 2a shows an example of FPN extracted from a blackbody image at 18 °C, using a Santa Barbara focal plane camera (model SBF125) operated at 157 Hz.

The third factor is a bad pixel, which can be an anomalous pixel behaving differently from the rest of the array. A bad pixel normally gives a black colour while a hot pixel gives white colour. Both of the type of pixel will not give any useful information and only spoils the image contrast. Figure 2b is an example with a bad pixel. The simple way to remove it is to replace the bad pixel with the average value of the neighbouring pixels.

3 Methodology

3.1 Square Pulse Thermography Experiment

Square pulse thermography is one of the active non-destructive tests that can be used to inspect and evaluate objects. The equipment that is used in this experiment is the specimen that represent the car's boot (aluminum alloy T6061 plate as shown

(a)   **(b)**

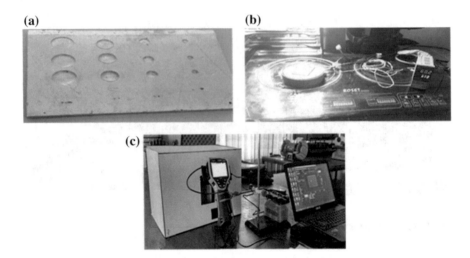

Fig. 3 **a** Aluminium alloy T6061 plate, **b** electric ring heater and **c** experiment setup

as in Fig. 3a) with applied defect, U5855A TrueIR Thermal Imager, box (cover of the specimen) electric ring heater and laptop (image processing and video recording).

Before the experiment begun, it was necessary to ensure that all the safety requirement was followed to reduce the risk of injury. In Fig. 3b, the initial temperature of the specimen must be recorded using both of the measurement tools, i.e. the thermographic camera and a contact thermometer. After that, the specimen was put on the heating source for 30 s. The regulator temperature was set to 60 °C. After the heating process, the specimen was placed on the specimen's stand and covered with the box. The camera was set to video mode for recording and the camera recorded the cooling process of the specimen. After the temperature of the specimen stabilized with the temperature of the room, the recorded video was saved to the SD card, and the video from the SD card was transferred to perform the analysis.

3.2 Image Processing

This section shows the image processing method that has been implemented during analyzing the raw data. The programming language python was used for the purpose. The integrated development environment (IDE) that was used are Spyder and the library that was utilized included Matplotlib, Numpy and OpenCV. The raw data in form of sequence of images was converted to gray scale during the first step. Taking this approach further results in a powerful technique for improving the appearance of images: the grayscale transform. The idea is to increase the contrast at pixel values of interest, at the expense of the pixel values we do not care about. This is done by defining the relative importance of each of the 0–255 possible pixel values. The more important the value, the greater its contrast is made in the displayed image.

The second step was to apply the discrete Fourier transform algorithm. Figure 4a show the raw image from the thermal imager and transform to the grayscale image in Fig. 4b. The output of the transformation represents the image in the Fourier or frequency domain (see Fig. 4c) while the input image is the spatial domain equivalent. In the Fourier domain image, each point represents a particular frequency contained in the spatial domain image. The Fourier transform is used in a wide range of applications, such as image analysis, image filtering, image reconstruction and image compression. In the magnitude spectrum, the rectangular window size of 60 × 60 filter is applied to the image. The result can be seen in Fig. 4d.

The third step is the edge detection using the Canny Edge Detection. Edges in images are sudden changes of discontinuities in an image. It also can be referring to significant transitions in an image. There are different types of edges which are the famous three including horizontal edges, vertical edges and diagonal edges. Most of the shape information of an image is enclosed in edges. So first we detect these edges in an image and by using these filters and then by enhancing those areas of image which contains edges, the sharpness of the image will be increased and the image will become clearer. Figure 4e shows the result after the using the Canny Edge Detection method.

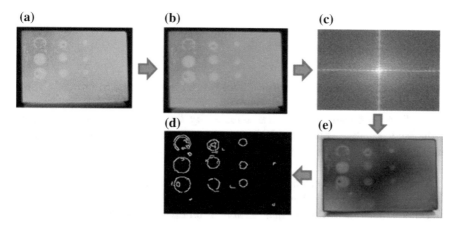

Fig. 4 **a** Raw image, **b** grayscale image, **c** frequency domain image, **d** image after fast fourier transform filter and **e** edge detection image

Table 1 Diameter estimation calculation

Defect on first's row			
68 pixel − 34 pixel	130 pixel − 104 pixel	183 pixel − 165 pixel	243 pixel − 234 pixel
=34 pixel	=26 pixel	=18 pixel	=9 pixel
=7.922 mm	=6.058 mm	=4.194 mm	=2.097 mm
Defect on second's row			
64 pixel − 28 pixel	123 pixel − 100 pixel	185 pixel − 168 pixel	241 pixel − 232 pixel
=36 pixel	=23 pixel	=17 pixel	=9 pixel
=8.388 mm	=5.825 mm	=3.961 mm	=1.08 mm
Defect on third's row			
73 pixel − 38 pixel	134 pixel − 108 pixel	189 pixel − 171 pixel	274 pixel − 264 pixel
=34 pixel	=26 pixel	=18 pixel	=10 pixel
=8.155 mm	=6.058 mm	=4.194 mm	=2.33 mm

3.3   Diameter Estimation Method

The diameter of the specimen can be determined using the ratio between the physical size and the pixel number in the thermogram of the tested specimen. The inspection surface and the infrared thermal imager is 0.3 m. With these arrangements, one pixel represented about 0.233 mm in the physical world. This number can be varied with a different setup. Table 1 shows the sample calculation to determine the diameter of the defect in the image. The estimation could not be accurate; we will elaborate further about this estimation in the discussion section.

4   Result and Discussion

4.1   Experiment Setup

During the experiment, a sequence of images has been captured in form of a video, where the length of the video was approximately 7.04 min. In the detection process, the surface temperature signal was affected by the influence of various factors including the ambient noise, original thermal equilibrium state, and uneven heating, etc. Thermal contrast is one of the indicators of infrared thermal wave nondestructive detection.

The aluminum plate has a low emissivity, near to 0.15. Emissivity is defined as the ratio of the energy radiated from a material's surface to that radiated from a blackbody. This means, more than 95% is in reflection from surrounding sources. To overcome this, black paint had been applied to the surface of the specimen to increases the emissivity. The reflected light from the surrounding had appeared on the image. Thus, this image is not valid to proceed to the next process. Even the heat distribution can be detected from the image, but all of the reading is not correct. During the experiment, two tools had been used for checking the temperature of the specimen, i.e. thermography and a contact thermometer. The contact thermometer had been used to validate the reading temperature from the thermography camera and if the temperature reading was different from the contact thermometer reading, the reading is inaccurate.

During the experiment, a box was also installed during the cooling process. This can reduce the light reflection and humidity from outside and less distract the experiment. During the experiment, the setting for ambient temperature for the camera must be suitably adjust suitable to the current ambient temperature. This can prevent the camera from misinterpret the temperature data during the recording. Figure 5 shows the result of an image with reflection.

The heating source that should be used in this experiment setup should be a halogen lamp. The reason why the heating source was changed is that, the halogen lamp had not enough power to heat up the specimen. The used halogen lamp only can transmit as a 200 W heating source. So, the halogen lamp was replaced by an electric heating device that can transmit higher power. The time used for transmitting was different because the halogen lamp emits the energy by radiation while the electric heating device emits the energy by using conduction. Radiation is much faster than conduction.

After all the restrictions, the best result from the experiment is shown in Fig. 6. The defect may appear on the raw data, but there is still noise that need to be remove and this can be done by using image processing. We can see the changes in colour of temperature during the cooling process. The cooling process takes until the temperature of the specimen is the same as the room temperature.

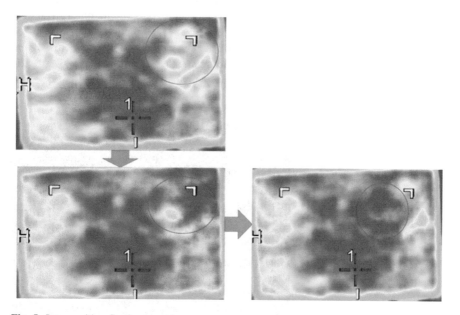

Fig. 5 Image with reflection

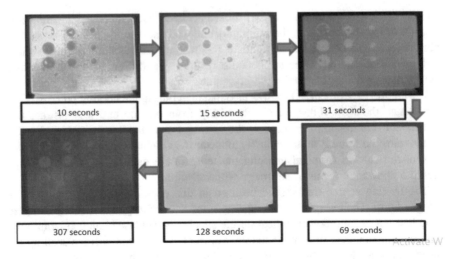

Fig. 6 Raw data image

4.2 Image Processing

Image processing is the way to enhance the visibility of the defect. This can be done by manipulating the image. The technique that was implemented in this project is the Fast Fourier transform. In order to remove the low frequency in the image we have to convert the image to the frequency domain then we applied the filter. After we applied the filter, we converted back the picture into the domain phase. The final result of the image processing is shown in Fig. 7. The smallest defect cannot be detected. This is because the defect was too small and there is still noise during the cooling process, so when the defect is too small, the camera failed to detect the defect. The circle of the defect is also not perfectly detected. There is still noise that distracts from the defect.

Fig. 7 Edge detection

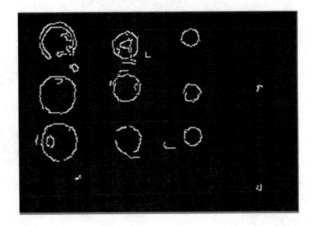

Table 2 Estimation error

Defect on first's row			
8 mm − 7.922 mm	6 mm − 6.058 mm	4 mm − 4.194 mm	2 mm − 2.097 mm
=0.078 mm	= −0.058 mm	= −0.194 mm	= −0.097 mm
=0.975% error	=0.120% error	=4.85% error	=4.85% error
Defect on second's row			
8 mm − 8.388 mm	6 mm − 5.825 mm	4 mm − 3.961 mm	2 mm − 2.097 mm
= −0.388 mm	=0.175 mm	=0.039 mm	= −0.097 mm
=4.85% error	=2.916% error	=0.975% error	=4.85% error
Defect on third's row			
8 mm − 8.155 mm	6 mm − 6.058 mm	4 mm − 4.194 mm	2 mm − 2.33 mm
= −0.155 mm	= −0.058 mm	= −0.194 mm	= −0.333 mm
=1.9375% error	=0.967% error	=4.85% error	=16.65% error

4.3 Results

During the estimation of the diameter, the percentage of error has been calculated. The estimation sometimes is impossible when dealing with small defects because the IR camera has problems to detect the presence of the defect. During estimation method, the diameter of the defect slightly differs from the actual reading. The average error based on the estimation of the diameter is less than 10%. Table 2 show the calculation of the error.

5 Conclusion

In conclusion, the objectives of this project have been successfully accomplished. Three objectives have been set out in order to solve the project process. A summary of the entire investigation is given as the following.

The first objective was to develop a square pulse thermography inspection system for automotive components. In order to achieve this objective, a literature review of the previous developments of thermography inspection systems was needed. By analyzing the raw data, the next process should be known. The sequence of data in form of images has been collected for the next phase. The experiment setup has been modified several times in order to solve problem for example, not enough heating power and their contradiction from the original problems statement.

The second objective was to analyse the effect of the defect magnitude on the surface temperature distribution. This objective can be achieved by analysis the raw data using the Spyder software. Manipulation of the raw image using several image processing techniques was needed in this task. The image processing technique include Fast Fourier Transform (FFT), grayscale transformation and edge detection using the canny technique.

The third objective was to determine a defect profile in metallic components using the square pulse thermography. To be able to determine the size of the defect, some calculation must be done in order to estimate the diameter of the defect. Then, the result in regards to the error must be stated so that future improvements can be done. Overall, thermography has demonstrated the ability to inspect automotive components in this research and has the potential to be an alternative way instead of ultrasonic and human's eyes merely.

Acknowledgements The authors gratefully acknowledge financial support for this work by the Universiti Kuala Lumpur Malaysian Spanish Institute (UniKL MSI) and System Engineering and Energy Laboratory (SEELab).

References

1. Brandão, P., Infante, V., Deus, A.M., Gholizadeh, S.: Thermo-mechanical modeling of a high pressure turbine blade of an airplane gas turbine engine: a review of non-destructive testing methods of composite materials. ScienceDirect Proc. Struct. Integrity **1**, 50–57 (2016). https://doi.org/10.1016/j.prostr.2016.02.008
2. Kumar, S., Vishwakarma, M., Akhilesh, P.: ScienceDirect advances and researches on non destructive testing: a review. Mater. Today: Proc. **5**(2), 3690–3698 (2018). https://doi.org/10.1016/j.matpr.2017.11.620
3. Niola, V., Quaremba, G., Amoresano, A.: A Study on Infrared Thermography Processed Trough the Wavelet Transform, pp. 57–62 (2009)
4. Theodorakeas, P., Cheilakou, E., Ftikou, E., Koui, M.: Passive and active infrared thermography: an overview of applications for the inspection of mosaic structures. J. Phys. Conf. Ser. **655**(1) (2015). https://doi.org/10.1088/1742-6596/655/1/012061
5. Maldague, X.: Introduction to NDT by active infrared thermography 1. Université Laval, Quebec City (2002)
6. Ibarra-castanedo, C., Bendada, A., Maldague, X.: Thermographic Image Processing for NDT. Université Laval, Quebec City (2007)
7. Ibarra-Castanedo, C., González, D., Klein, M., Pilla, M., Vallerand, S., Maldague, X.: Infrared image processing and data analysis. Infrared Phys. Technol. **46**(1–2 SPEC. ISS.), 75–83 (2004). https://doi.org/10.1016/j.infrared.2004.03.011

Deep Neural Network Modeling for Metallic Component Defects Using the Finite Element Model

Liyana Isamail, Nor Liyana Maskuri, Neil Jeremy Isip, Siti Farhana Lokman and Muhamad Husaini Abu Bakar

Abstract Nowadays, quality assurance is important for companies that are manufacturing components for various uses especially in the automotive industries. However, the inspection systems for determining the quality of these components are usually done by human workers which sometimes lead to inconsistencies. In order to counter this issue, an image classification-based technique using a Convolutional Neural Network (CNN) algorithm is introduced in this paper. The CNN provides a better approach in learning a feature data hierarchy to distinguish among samples of the defect and non-defect data represented as colored images. The process involves: (i) Region extraction using the finite element model, (ii) Formulate the model using a deep learning-based CNN algorithm, (iii) Defect detection. Four sets of metal dataset were used to train the model and to verify the accuracy and stability of the proposed method. The results demonstrated that the proposed CNN model can predict defects and non-defects data with the accuracy of 100%, precision of 99%, recall of 100%, and F1-score of 100%. Based on the experimental result, the proposed model is expected to be promising due to its robustness which can be used to detect defects in an online detection in ensuring quality manufacturing components.

Keywords Neural network · Convolution neural network · Finite element model · Deep learning · Metallic defect

L. Isamail · N. L. Maskuri · N. J. Isip · S. F. Lokman · M. H. Abu Bakar (✉)
System Engineering and Energy Laboratory, University Kuala Lumpur-Malaysian Spanish Institute, Kulim Hi-Tech Park, 09000 Kulim, Kedah, Malaysia
e-mail: muhamadhusaini@unikl.edu.my

L. Isamail
e-mail: liyanaisamail@gmail.com

N. L. Maskuri
e-mail: nliyana.maskuri@s.unikl.edu.my

N. J. Isip
e-mail: jeremy_neil1396@yahoo.com

S. F. Lokman
e-mail: sisfaeez@gmail.com

M. H. Abu Bakar et al. (eds.), *Progress in Engineering Technology*,
Advanced Structured Materials 119, https://doi.org/10.1007/978-3-030-28505-0_23

1 Introduction

Machine learning (ML) has recently gained in popularity, spurred by well-publicized advances like deep learning and widespread commercial interest in big data analytics. Machine learning is a method of data analysis that automates analytical model building. Using various algorithms that iteratively learn from data, machine learning allows computers to find hidden insights without being explicitly programmed on where to look. It has a basic structure to process and produce outputs [1]. The surge of interest in machine learning is due to data mining factor where things like growing volumes and varieties of available data are accessible via computational processing that is cheaper and more powerful, and affordable data storage. All these things mean it is possible to quickly and automatically produce models that can analyze bigger, more complex data and deliver faster, more accurate results even on a very large scale.

Machine learning algorithms can be categorized to supervised learning, unsupervised learning and deep learning. Supervised machine learning techniques is a training set of examples which automatically learn a model of the relationship between a set of descriptive feature and a target feature, the algorithm generalizes to respond correctly to all possible inputs [2, 3]. In unsupervised learning, the data obtained not having label and it can be due to the unavailability of funds to pay for manual labelling or the inherit nature of the data itself, but the main idea is to find a hidden structure in the data [4]. Deep learning is a subcategory of machine learning in which multiple-layered networks are used to assess complex patterns within the raw imaging input data. The multiple-layered network which is responsible for the 'learning task' is called neural network. The benefit of machine learning is to give high-value predictions and to allow humans to choose better decisions and smart actions in real time. The typical neural network usually consists of three-layer types which are Convolutional Layers, Max-Pool Layers and Dense Layers [5] and the typical architecture is a sequence of feed forward layers implementing convolutional filters and pooling layers [6]. Figure 1 shows the standard model of a neural network.

Although the majority of the quality inspection technologies of vehicles are using the NDT technology, the quality control process for automotive industries nowadays however, is still carried out using manual detection and subjective evaluation by experts, known as check-men [7]. Thus, the detection and evaluation are still not completely reliable due to errors performed by humans. Therefore, to counter the problems faced nowadays, the use of the NDT technology with the help of deep neural networks provide a better output on the detection of the defects in automotive components especially in metallic materials.

Again, automotive inspection systems nowadays use the availability of non-destructive (NDT) technologies to fully inspect a vehicle. Example of NDT—based technologies are the use of ultrasonic testing, liquid penetrant testing and etc. These quality control processes are still carried out by using manual detection and subjective evaluation by experts, known as check-men [8]. Thus, this method is

Hidden

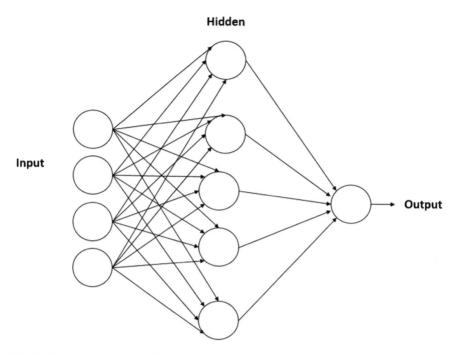

Fig. 1 Neural network model [1]

only available for experts due to difficulties in data interpretation. This results to a limited acquisition of the inspection into few samples.

Besides that, all these technologies are used to collect real-time data from machines and therefore massive data are obtained after long-time operation of the machine. However, according to [8], the reports of the inspection are compiled by humans which are affected by human errors, moreover, only the most important of these are reported and saved to databases, resulting in the loss of valuable information. But, with the use of neural networks, we can store large input data to the system which are then able to produce outputs which we required. On top of that, the large input data helps the system to be more accurate in producing the outputs therefore no valuable data are wasted.

On top of that, conventional computers use an algorithmic approach to solve a problem. Meaning that the computer follows a set of instructions in order to solve a problem. However, neural networks can learn by example. They cannot be programmed to perform a specific task which proves that it can automatically adapt based on the data fed. The examples must be selected carefully otherwise useful time is wasted or even worse the network might not be functioning correctly. Geoffrey Hinton at Google boasts, "Deep Learning is an algorithm which has no theoretical limitations of what it can learn; the more data you give and the more computational time you provide, the better it is" [9].

This paper focuses on the modelling of a deep learning neural network model to be able to predict the presence of defects on a metallic component. Moreover, the proposed project is divided into two stages. The first stage will cover the collection of datasets obtained from the finite element method in the form of images. The second stage covers the modelling of the neural network model using a convolutional neural network. On top of that, the classification algorithm is used to predict the presence of defects on a specific specimen.

2 Methodology

2.1 Project Process

The proposed project consists of three stages. The project starts off with the 3D modelling of the specimen using the Abaqus CAE software. Various defects are modelled on the specimen to provide different types of data in the form of images. The specimen is then undergone a transient heat analysis to provide a temperature contour model which is needed in the dataset collection of images of defects as well as for non-defects. Every sequence of image in the analysis throughout the 30 s of the heat analysis will be taken as a dataset for both defects and non-defects specimens. About 256 of images for both defects and non-defects data are obtained. These data are then categorized as training data for the neural network. Moreover, another new set of data of about 56 images of defects and non-defects specimen are created to provide a testing data for the neural network. After that, these data will then be fed into the convolutional neural network with the use of classification algorithm to classify images with defects as well as non-defects.

2.2 Finite Element Analysis

The computer-aided design (CAD) of the specimen in Abaqus CAE is based on an actual experimental setup for a thermography test. Figure 2 shows the dimensions of the CAD design of the specimen.

The 3D modeling of the specimen is then constructed based on the dimensions used. Figures 3 and 4 show the actual specimen used on a thermography test and the CAD design of the specimen, respectively. Transient thermal analysis determines temperatures and other thermal quantities that vary over time. A transient thermal analysis follows basically the same procedures as a steady-state thermal analysis. The main difference is that most applied loads in a transient analysis are functions of time. Example of heat transfer applications and heat transfer problems that involve transient thermal analysis are nozzles, engine blocks, piping systems, pressure vessels, etc.

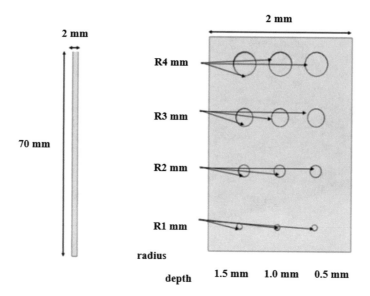

Fig. 2 Dimension of the specimen

Fig. 3 Actual specimen used in the thermography test

Based on the simulation, the transient thermal analysis starts with the identification of the properties of the material. For a transient thermal analysis in
Abaqus CAE, only the density (ρ), thermal conductivity (k) and specific heat
capacity (C) are used in order to simulate the heat transfer simulation. Table 1
shows the material properties used for the finite element analysis.

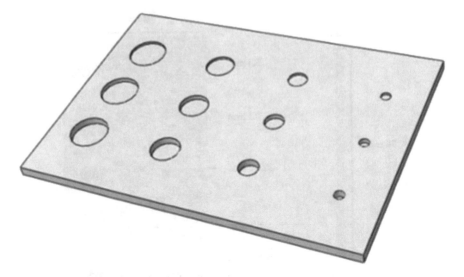

Fig. 4 CAD design of the specimen

Table 1 Properties of the specimen (Retrieved from ASM Aerospace Specification Metals Inc.)	Material	Aluminium 6061–T6
	Mass	17.2 g
	Mass Density, ρ	2457.1429 kg/m^3
	Specific heat capacity, c	895 J/kg °C
	Thermal Conductivity, k	167 W/mK

Next, the boundary condition as well as the loads to be applied on the specimen are important in replicating an actual experimental setup for a simulation. For thermal analysis, a 5000 W/m^2 of heat flux is applied to the top of the specimen where the defects were present with a free boundary condition on the bottom to allow temperature to increase freely as is setup in the actual experiment. The time period for the transient heat transfer is set to 30 s.

Once the simulation is finished, the result can be displayed in the form of an image or animations as well in a sequence of images. In this phase, the image of the specimen was taken from the backside where the defects are not noticeable. Every image from every transient heat analysis from the FEM was captured in sequence for the period of 30 s. This sequence of images represents the heat distribution on the metal specimen over time. Figure 5 shows examples of the image taken from the software.

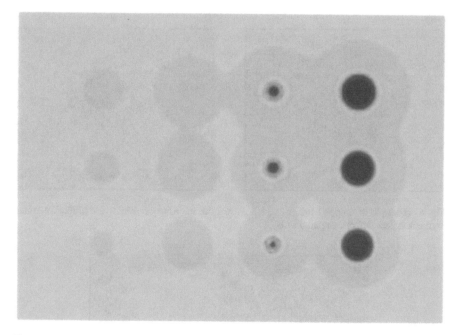

Fig. 5 Example of temperature contour mode

2.3 Deep Neural Network Modelling

The modelling of the convolutional neural network model used a high-level neural networks application programming interface (API) called Keras which is written in the Python programming language. The environment where the preparation of the programming was done is the Spyder Integrated Development Interface (IDE). Before feeding the dataset to the neural networks, each of the defect features for every image was cropped into specific dimensions to provide better recognition of defects features for the convolutional neural network model. Figure 6 shows the example of cropped images of defect features.

3 Result and Discussion

3.1 Performance Analysis of Proposed Convolution Neural Network Model

Based on the confusion matrix generated by the current proposed model, it consists of 4 different combinations of predicted and true labels. The true labels consist of the actual 'Defect' and 'Non-Defect' label whereas the predicted label also shows the

(a)   **(b)**

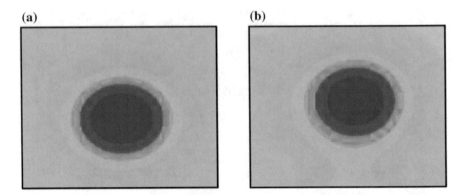

Fig. 6 Example of cropped images which shows the intensity of defects for different surface. Figure **a** is less defect compared to **b**

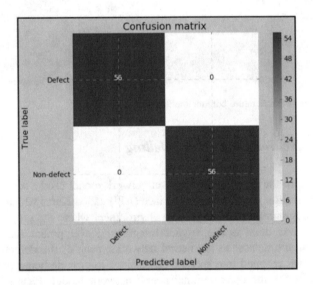

Fig. 7 Confusion matrix of current neural network model

same classification. Besides that, the number 56 represents the number of images for the brand-new set of data for each of the defects and non-defects specimen as stated above. The results show that the model successfully predicts the correct outcome for each of the 112 brand-new set of images. The model predicted that 56 images were defected and another 56 images were non-defected. This shows that the proposed model shows a promising result classifying between defected and non-defected specimens. Figure 7 shows the confusion matrix of the current proposed model.

Furthermore, we can then proceed to evaluate the performance metrics of the model. Table 2 shows two classification classes which are defects and non-defects

Table 2 Performance matrix of current model

	Precision	Recall	F1-Score	Support
1 (Defect)	1.00	1.00	1.00	56
2 (Non-defect)	1.00	1.00	1.00	56

which are represented by 1 and 0 respectively. The performance metrics used in this table are Precision, Recall, F1-score and Support.

Precision metric is the ratio of correctly predicted positive observations to the total predicted positive observations. Based on the table, it shows the ratio value of 1.00 which corresponds to the confusion matrix predicting the correct condition for the 56 images of defects and non-defects. Recall is the ratio of correctly predicted positive observations to the all observations in the actual class. By referring to the table, the recall metric shows a value of 1.00 which corresponds to the same condition as in the Precision metrics.

Moreover, F1-Score shows the weighted average of Precision and Recall. Hence, this score takes both false positives and false negatives into account. The F1-Score is usually more useful than accuracy, especially when dealing with uneven class distributions. Accuracy works best if false positives and false negatives have similar cost. If the cost of false positive and false negatives is very different, it is better to look at both Precision and Recall. On top of that, the F1-Score shows a value of 1.00 which indicates that it is at the best score whereas the worst score will have the value 0.

Finally, the Support is the number of occurrences of each class which is 56 for defected images and another 56 for non-defected images. Table 2 shows the performance metrics for the proposed neural network model. Again, a graph of the training loss and validation loss with respect to the number of epochs used can be plotted to visually show the performance of the proposed model. Epoch is the complete pass through as given dataset. Training loss is the error on the training set of data whereas the validation loss is the error after running the validation set of data through the trained network. Based on the Fig. 8, it shows that the training starts at about 0.38 at epoch = 0. Then, it starts to gradually decrease as the number of epochs increases. This shows that the model starts to learn the feature of each images as the epoch increases. This then leads to the training loss eventually drops down to zero. This indicates that the model successfully learns the features of each image fed.

Next, we look at the validation loss of the model. Referring to the Fig. 8, the line for validation loss are untraceable due to the scale of the graph. Therefore, we refer to Fig. 9, which shows that the loss starts at 0.0003 at epoch = 0 and then stabilizes at epoch = 1. Then, an increase of loss at 0.00095 at epoch = 5 which then stabilizes at epoch = 7. The sudden increase in the validation loss is due to the model was tested on recognizing different images to classify between defect and non-defect specimens. However, as the epoch increases, it shows that the model

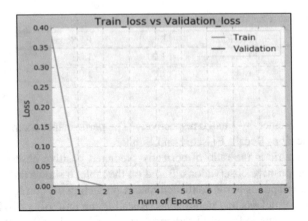

Fig. 8 Cross validation between train loss and validation loss

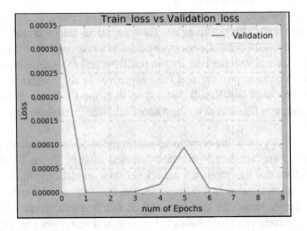

Fig. 9 Graph of validation loss

successfully recognized the feature as well as predict the correct outcome. Overall, we can ensure that the learning rate of the model is not overfitting as the validation loss is lower than the training loss.

In addition, we can also look at the graph of training and validation accuracy to be able to depend on the proposed model. Based on Fig. 10 on the training accuracy and validation accuracy, it shows that the training accuracy for the model increases as the epoch increases which eventually lead to an accuracy of the value 1.00. This shows that the model learns progressively better as the epoch increase. Besides that, the validation accuracy in Fig. 11 shows an accuracy of 1.00 which proves that the model is able to classify a new set of images correctly after training stages.

Fig. 10 Cross validation of train accuracy and validation accuracy. The validation accuracy is overlapping with the train accuracy

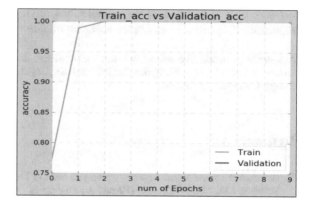

Fig. 11 Graph of validation accuracy

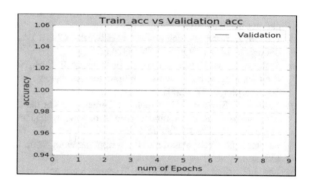

4   Conclusion

As a conclusion, quality assurance on the product manufactured in a manufacturing industry is important in order to provide an excellent quality to customers. The implementation of neural networks in the quality inspection system of a specific component, for example, a metallic automotive component, will provide a better performance in terms of detecting defects in a component. This is due to the increased reliability of the neural network to easily interpret data as well as hinder major errors in providing desired results in predicting the outcome. So, the proposed project is to provide a better approach in detecting defects in a metallic component using convolutional neural networks. Based on the result obtained, the model shows good result in predicting the correct outcome for different images. Moreover, the loss and accuracy graph as well as the confusion matrix provide a better understanding on the performance of the model. The graph indicates that the model is not overfitting which means it has good performance on training data as well as generalizing different data. Furthermore, the results demonstrated that the proposed CNN model can predict defects and non-defects data with the accuracy of 100%, precision of 99%, recall of 100%, and F1-score of 100%. So, the neural network proved to be more dependable than relying on human on specific tasks especially in quality assurance on ensuring quality manufacturing components.

Acknowledgements The authors gratefully acknowledge financial support for this work by the Universiti Kuala Lumpur Malaysian Spanish Institute (UniKL MSI) and System Engineering and Energy Laboratory (SEELab).

References

1. Parker, M.: Digital signal 101: everything you need to know. (2017)
2. Kelleher, J.D., Mac Namee, B., D'Arcy, A.: Fundamentals of machine learning for predictive data analytics. (2015)
3. Marsland, S.: Machine learning & pattern recognition series chapman & Hall/CRC. (2009)
4. Mohammed, M., Muhammad, Khan, M.B., & Bashier, E.B.M.: Machine learning : algorithms and applications. CRC Press, (2017)
5. Heaton, J.: Deep learning and neural networks. CEUR Workshop Proceedings. (2015). https://doi.org/10.1017/CBO9781107415324.004
6. Mohsen, H., El-Dahshan, E.-S.A., El-Horbaty, E.-S.M., Salem, A.-B.M.: Classification using deep learning neural networks for brain tumors. Future. Comput. Inf. J. (2017). https://doi.org/10.1016/j.fcij.2017.12.001
7. Molina, J., Solanes, J.E., Arnal, L., Tornero, J.: On the detection of defects on specular car body surfaces. Rob. Computer-Integrated. Manuf. **48**(January), 263–278. (2017). https://doi.org/10.1016/j.rcim.2017.04.009
8. Ciaburro G., Venkateswaran B.: Neural network with R. Birmingham. United Kingdom, (2017)
9. Lee J.H., Shin J., Realff M.J.: Machine learning: overview of the recent progresses and implications for the process systems engineering field. (2018)

Printed in the United States
By Bookmasters